Technopolitical Mediation

Postphenomenology and the Philosophy of Technology

Editor-in-Chief: Robert Rosenberger
Executive Editor: Peter-Paul Verbeek

As technologies continue to advance, they correspondingly continue to make fundamental changes to our lives. Technological changes have effects on everything from our understandings of ethics, politics, and communication, to gender, science, and selfhood. Philosophical reflection on technology can help draw out and analyze the nature of these changes, and help us to understand both the broad patterns of technological effects and the concrete details. The purpose of this series is to provide a publication outlet for the field of philosophy of technology in general, and the school of thought called "postphenomenology" in particular. The field of philosophy of technology applies insights from the history of philosophy to current issues in technology and reflects on how technological developments change our understanding of philosophical issues. Postphenomenology is the name of an emerging research perspective used by a growing international and interdisciplinary group of scholars. This perspective utilizes insights from the philosophical tradition of phenomenology to analyze human relationships with technologies, and it also integrates philosophical commitments of the American pragmatist tradition of thought.

Recent Titles in This Series

Technopolitical Mediation: An Arendtian Approach to Political Philosophy of Technology by Melis Bas
Postphenomenology and Technologies within Educational Settings, edited by Markus Bohlmann and Patrizia Breil
Modernity and Technology: A Philosophical Investigation of Martin Heidegger and Bruno Latour by Søren Riis
Moral Hermeneutics and Technology: Making Moral Sense through Human-Technology-World Relations by Olya Kudina
Postphenomenology and Imaging: How to Read Technology, edited by Samantha J. Fried and Robert Rosenberger
Postphenomenology and Architecture: Human Technology Relations in the Built Environment, edited by Lars Botin and Inger Berling Hyams
Technology and Anarchy: A Reading of Our Era by Simona Chiodo
How Scientific Instruments Speak: Postphenomenology and Technological Mediations in Neuroscientific Practice by Bas de Boer
Feedback Loops: Pragmatism about Science & Technology, edited by Andrew Wells Garnar and Ashley Shew

Technopolitical Mediation

An Arendtian Approach to Political Philosophy of Technology

Melis Bas

BLOOMSBURY ACADEMIC
NEW YORK • LONDON • OXFORD • NEW DELHI • SYDNEY

BLOOMSBURY ACADEMIC
Bloomsbury Publishing Inc, 1359 Broadway, New York, NY 10018, USA
Bloomsbury Publishing Plc, 50 Bedford Square, London, WC1B 3DP, UK
Bloomsbury Publishing Ireland, 29 Earlsfort Terrace, Dublin 2, D02 AY28, Ireland

BLOOMSBURY, BLOOMSBURY ACADEMIC and the Diana logo are trademarks of Bloomsbury Publishing Plc

First published in the United States of America 2026

Cover design by Becca Huguley
Cover image © iStock.com/imagedepotpro

Library of Congress Cataloging-in-Publication Data is available

ISBN: HB: 978-1-6669-5716-7
ePDF: 979-8-216-25957-2
eBook: 978-1-9787-6541-2

Series: Postphenomenology and the Philosophy of Technology

Typeset by Integra Software Services Pvt. Ltd.

For product safety-related questions, contact productsafety@bloomsbury.com.

Contents

List of Figures vii

Introduction: Technology, Power, and Politics 1
Politics of Technology 2
Technology and Power 3
Questioning Power 6
A Viable Alternative: Issues and *Dingpolitik* 8
Hannah Arendt: A Guide for a New Political Mediation Theory 12
Technological Mediation 14
Outline 16

Part 1

1 Politics as Interaction: Hannah Arendt and Political Theory 23
Plato's Transformation of the Polis 24
Vita Activa 28
Features of Politics 34
Conclusion 51

2 Arendt, Phenomenology, and Technology 55
Hermeneutic Phenomenology of the Political 56
Common World and Worldliness 74
Technology in Arendt's Work 86
Cultural Products of *Homo Faber* 100
Conclusion 109

Part 2

3 Technological Mediation of Common Sense 113
From Phenomenology to Postphenomenology 114
Empirical Explorations 129

Gezi Park and Technological Mediation of Common Sense 131
Conclusion 150

4 Technological Mediation of Intersubjectivity 153
Political Subjectivity and Intersubjectivity 155
The Polis-Making of Gezi Park 175
Technological Mediation of Intersubjectivity 185
Technopolitical Mediation 192
Technopolitical Mediation on Social Media 196
Conclusion 203

Conclusion 207
Revisiting the Moses Bridges with Technopolitical Mediation 208

References 214
Index 226

Figures

1	Technological Mediation Theory Scheme	124
2	The Hermeneutic Lemniscate	126
3	The Hermeneutic Lemniscate (the first move)	148
4	The Hermeneutic Lemniscate (the second move)	194

Introduction: Technology, Power, and Politics

It became clear on November 6, 2024, that the former president, Donald Trump, was going to be the forty-seventh president of the United States. The days leading up to the election were chaotic, and it has been argued that they were full of disinformation (Cohen et al., 2024; Wendling, 2024; West, 2024). Some examples of disinformation included a video circulating online claiming that vice presidential candidate Tim Walz had attacked a former student of his (Gilbert, 2024), an image of Kamala Harris hugging convicted sex offender Jeffrey Epstein on a beach (Eifert, 2024), and images of Donald Trump walking through a flooded street (Hudnall, 2024) and calling federal prosecutor Jack Smith a loony leftist (McCreary, 2024). What is a common feature in all these images and videos is that they were all fabricated, through either artificial intelligence or other technological means. The political landscape nowadays clearly signals one thing: technology and politics are already inseparable. Clearly, technological choices lead to political consequences.

For a long time, philosophers of technology conceptualized the political significance of technology as a matter of autonomy, specifically whether it is humanity or technology that has autonomy. Autonomy can essentially be seen as self-governance, in which one's behaviors are dictated by internal factors, wants, circumstances, and traits that represent one's own self rather than being imposed from elsewhere (Christman, 2020). Especially where politics is concerned, the concept of autonomy has been coupled with the notion of power. By centering power as the main dynamic in politics, such approaches overlook the broader political significance of human–technology relations, especially how technologies shape our capacity to form political communities and interpret shared realities. In this book, I show that the political philosophy of technology needs to move beyond an analysis of how technologies exert power over people and focus on analyzing how they facilitate interaction and shape the formation of political communities. I develop a framework that moves beyond the notion of power as the focus of the politics of technology by integrating Arendt's political theory with technological mediation theory. By emphasizing Arendt's understanding of political hermeneutics and intersubjectivity, this approach developed in this book reframes technology's role as a mediator of collective interpretation and collective interaction, thereby enabling new forms of political coexistence.

In this introductory chapter, I provide a brief outline of the political philosophy of technology through an examination of three major approaches in the field, represented by the work of philosophers Langdon Winner, Andrew Feenberg, and Bruno Latour. In doing so, I first examine the concept of power that informs major approaches in the political philosophy of technology. By exposing the limits of existing approaches, I problematize the basis of politics as power and propose an alternative to this existing account: politics as interaction. Based on this proposal, I then explore the approach developed in this book.

Politics of Technology

In his famous article "Do Artifacts Have Politics?" (Winner, 1980), Langdon Winner gives three examples of how technologies can have political or social consequences. His most famous example is the low-hanging bridges on Long Island, New York, designed by Robert Moses (Winner, 1980). Relying on the biography of Moses by Robert A. Caro, Winner argues that Moses, one of the most influential architects in New York from the 1920s to the 1970s, built the overpasses on Long Island so low specifically to prevent public transportation vehicles such as buses (which are around nine feet tall) from passing below them (Winner, 1980). According to Winner, Moses inscribed his ideas about social class and racial prejudice in the artifacts he designed (Winner, 1980). The overpasses were built in such a way that the "automobile-owning" white upper- or middle-class citizens were able to move to specific locations on Long Island for recreation (Winner, 1980). Meanwhile, poor people who needed to use public transportation (and who were predominantly Black) could not travel about as easily, because the overpasses prevented the buses from reaching the beach (Winner, 1980). For Winner, this design choice was made deliberately to achieve a "particular social effect" (Winner, 1980, p. 123).

On the basis of this and other examples, Winner claims that technologies can be inherently political; they can embody ideologies and "specific forms of power and authority" (Winner, 1980, p. 123). Moses' bridges are "political," according to Winner, because they carry out the racially biased ideology of their designer (Winner, 1980). Since it was published, Winner's analysis of the "politics of artifacts" has been extensively debated and challenged. Some authors have raised questions about the accuracy and usability of the example: Joerges, for instance, argues that bus timetables show the overpasses did not really obstruct all bus traffic (Joerges, 1999), and Woolgar and Cooper even characterize Winner's example as an "urban legend" (Woolgar & Cooper, 1999, p. 439). What has not been challenged in the various analyses and critiques, though, is that the main political issue at stake is the potential power of the overpasses to obstruct bus traffic and, as a result, to discriminate against specific groups of people. The political dimension of technological artifacts, therefore, is primarily analyzed in terms of *power*, which resonates well with the definition of politics as "arrangements of power and authority in human associations as well as the activities that take place within those arrangements" (Winner, 1980, p. 123). In addition, the concept of power at the heart of Winner's definition of politics is concerned with domination. In other words, in analyzing the political dimension of human–technology relations, Winner is concerned with who has power over whom.

Technology and Power

Like Winner, Andrew Feenberg, another well-established philosopher of technology, defines technology as "one of the major sources of public power in modern societies" (Feenberg, 1992, p. 301). Feenberg conceptualizes the relationship between humans and technologies as a struggle for power and domination. According to Feenberg, the most fundamental distinction that can be made between technical actors is in the form of dominance (Feenberg, 1999, p. x). For him, dominance or subordinacy rhetoric is visible in technical systems, and humans—in relation to technology—have a subordinate position. The moment we humans realize that we are somewhat subordinated to technical systems, argues Feenberg, we can begin to intervene in the design process of technologies and thus affect the direction of the development of modern society (Feenberg, 1999). Technologies, according to Feenberg, are not purely efficiency-driven devices; rather, they are determined by design choices. Thus, the more frequently regular citizens intervene in the design process of technologies, the more they will be able to direct technologies toward their understanding of meaningful life (Feenberg, 1999, p. xiv). Drawing on his theory of critical constructivism (Feenberg, 1999, 2010), Feenberg provides a framework to begin this intervention in the design process of technologies and thus affect the direction of society's development. Feenberg calls this process of giving direction to technologies "democratic interventions." Through democratic interventions, Feenberg believes the asymmetrical division of power between technical actors (humans and technologies) can be eliminated, and thus humans can be actively involved in shaping the direction of technology (Feenberg, 2010).

To better illustrate what critical constructivism entails, Feenberg offers a case study, namely that of Minitel. Minitel was the French national videotex online system, which was accessible through mainland telephone lines (Cats-Baril & Jelassi, 1994). The evolution of this technology amply demonstrates what Feenberg calls "democratic interventions" (Feenberg, 1999, p. 90) in technical design. In the mid-1980s, the French telecommunications operator decided to incorporate computer conferencing into its own Minitel system and distributed millions of Minitels. These terminals were made to imitate the experience of a regular domestic telephone and were aimed at providinga connection to information services (Feenberg, 1999). Aimed at providing a network for users to talk to each other, the terminals were distributed for free, and users were not required to have any knowledge of computers (Feenberg, 2010). However, although its aim was the distribution of information, the Minitel was soon redefined by its users, who employed it for anonymous online chatting in search of amusement, companionship, and sex (Feenberg, 1992). The users changed the purpose of the Minitel system to "the very opposite of the rationalistic project for which it was originally designed" (Feenberg, 2010, p. 15). Designed to solve the problem of the distribution of information, Minitel was addressed by its users to a different problem: human communication. Through design changes, the new interpretation of the technology was subsequently incorporated into the structure of the Minitel system, ultimately causing a change in its very definition (Feenberg, 1999).

According to Feenberg, the Minitel case portrays an excellent example of democratic interventions in technology. The actions of some hackers caused a change in the meaning of the Minitel system itself because millions of users adopted the same way of using the artifact. For Feenberg, this democratic intervention was significant and successful because of the vast number of users who participated in the act. Additionally, Feenberg points out that this technology was introduced to pursue efficiency-based interests by businesses. However, after user interventions, the sociotechnical structure of Minitel transformed from being a top-down system where people connected individually to central hubs to a social system where everyone was able to connect with one another for personal communication (Feenberg, 2008).

Through the case of Minitel and many other examples, Feenberg shows that "resistance is the practice when those threatened by technology control technology" (Feenberg, 1995, p. 207). Thus, critical constructivism is a way to analyze the forms of resistance to the conditioning of technological arrangements that limit the scope of human activity. A technological design may be intended to serve a specific purpose, but human beings intervene and give new meanings to the technology via these democratic interventions, which can vary in form: voicing public protest, inventing a novel use of the technology in question, or engaging in participatory design (Rao et al., 2015). Feenberg claims he is not suggesting that technologies and humans are opposing one another, but—with case studies such as Minitel—he emphasizes that the strategy to counter the rule of technocracy should be the redistribution of power by means of democratic interventions (Feenberg, 1999).

Over the years, Feenberg's approach has garnered both praise (Cohen, 1994; Doppelt, 2001; Jandrić, 2019) and criticism (Rao et al., 2015; Veak, 2000; Verbeek, 2013). For example, according to Verbeek, Feenberg positions the fundamental human–technology relationship as a struggle that can end in either oppression or liberation. While Feenberg aims to avoid suggesting an opposition between humans and technologies, he still ultimately formulates human–technology relations as a struggle that requires resistance on the part of humans (Verbeek, 2013). According to Verbeek, this approach is futile, since it adheres to a reificatory concept of power and causes Feenberg to miss the interdependent relationship between technology and humans in which they coshape each other's existences (Verbeek, 2013). Verbeek's theory of technological mediation considers humans and technologies as intertwined and equally influential in the production of a good life (Verbeek, 2011). In reply to Verbeek's criticism, Feenberg (2013) claims that, like Verbeek, he is also interested in exploring new approaches to analyzing how technological mediation can shape the good life. Thus, for Feenberg, his research focus is similar to Verbeek's. Struggle is undoubtedly evident in Feenberg's work, but according to Feenberg, this struggle is mainly between social groups with diverse interests within the technical system rather than between technology and society (Feenberg, 2013).

Questioning Power

It can be argued that Winner and Feenberg understand politics exclusively in terms of *power relations*, instigated by technologies that affect human actions. Yet, even though power is certainly an essential concept in political philosophy, it cannot sufficiently account for what politics is and can be. The framework of power is not the only perspective on the politics of technology—and ultimately on the question of autonomy. How should the concept of power be understood as it is related to the politics of technology?

There is a vast literature on power, and in the past several decades, the concept of power has been strongly conceptualized through two different dimensions: *power over* and *power to* (Göhler, 2009). For Haugaard, these two contrasting views dictate the debate (Haugaard, 2012), one of which focuses on "power as *domination* which is largely characterized by *power over*" (Haugaard, 2012, p. 33). According to Göhler, *power to* is characterized "either as a capacity or as empowerment" (Göhler, 2009, p. 34).

Haugaard (2012) refers to Dahl (1957) and argues that *power over* entails an agent A having the ability to compel another agent B to "do something which B would not otherwise have done" (Dahl, 1957, pp. 202–3). According to Haugaard, this conceptualization of power reflects the classical Weberian notion of power, normatively interpreted as domination (Haugaard, 2012; see also Göhler, 2009). This power relation is always at the expense of the one at whom the power is directed. According to Göhler,

> Exercising *power over* within a social relation always produces a negative result for those subjected to it, because it narrows their field of action. This is the case regardless of the possibly noble intentions or positive outcomes of the exercise of power. A's autonomy within a power relationship necessarily means correspondingly less power for B.
>
> (Göhler, 2009, p. 28)

Winner and Feenberg's definition of politics is informed by this concept of power. Winner, for example, places *power over* at the core of his analysis of Moses' bridges: they exert power over those who want to use the road. A concept of politics that adheres to this definition of power is concerned with which actor in its relation to other actors will ultimately achieve domination.

While this notion of politics as *power over* is valuable, it is also limited; it focuses on how an agent can enforce its choices or vision of a good life on others, leaving out of sight how members of a society, through their interactions, communicate and discover different goals and ideals concerning how to live. In *power over*, there is no room for a multiplicity of perspectives that can reinforce one another; there is only one overpowering perspective. In addition, equating politics with the notion of *power over* limits the political analysis of human–technology relations to oppression versus resistance (Verbeek, 2013), thus losing sight of other ways in which technologies are politically relevant, for instance, by giving rise to public debate or by helping shape public spaces.

The other perspective on power, as mentioned above, is *power to*. According to Göhler, *power to* is conceptualized in the literature in two ways. On the one hand, it denotes the capacity to achieve something; on the other hand, it means generating the autonomous empowerment of an individual or a group (Göhler, 2009, p. 34). Göhler argues that, in the first conceptualization, *power to* is a potential, which means that its capacity can only be assessed when it is actualized (Göhler, 2009). In that sense, power is latent. In the second conceptualization, *power to* denotes an actual situation. Göhler argues that *power to* as empowerment either exists or does not exist (Göhler, 2009, p. 34). In this conceptualization, power is understood as a self-referential relation, not referring to one individual, but to a group and therefore to a community of individuals (Göhler, 2009, p. 32). Power in this sense is not only a capacity but also empowerment, people gaining the ability to act autonomously, and it is realized power through communication (Göhler, 2009, p. 32). However, Göhler adds that this form of power "does not exist without being permanently actualized" (Göhler, 2009, p. 35). Power in this sense is conceptualized as productive and generated through reciprocal interaction.

As can be seen, the differing concepts of power have one point in common: the fact that they rest on relationality—in other words, on *interaction*. *Power over* denotes a social situation where one has power over someone else (Raffnsøe, 2013). *Power to*, or power as empowerment, creates mutually dependent relationships: the power of A is strengthened by the power of B and vice versa (Goehler, 2000). Grasping the inherent relationality of power allows the debate to move beyond the limitations of politics perceived as *power over*. However, even shifting from *power over* to *power to* as the focus of politics does not address the main subject of politics itself—the way in which human beings relate to one another and how these relations constitute their reality. Power, either in the form of *power to* or in the form of *power over*, always implies and presupposes another source of power. In that sense, power can only be conceived of within a relation, since one would always need another to articulate or to understand what power is.

It thus can be argued that to understand the notion of politics in the work of Winner and Feenberg, it is first necessary to thematize the dimension of interaction: even if one intended to understand politics only in terms of *power over*, one would still first need to grasp the basis of the power relation, which is interaction, since interaction is already inscribed in the notion of *power over*. The interaction might end in one agent overpowering another agent; however, as mentioned, for this result to be achieved, interaction—and the role of technology in this interaction—must first be thematized. Only when the focus is on an investigation of how humans live together can such a concept of politics facilitate a clear grasp of the role of technologies in this living together.

A viable alternative to Winner and Feenberg's *power over*-based politics of technology is Bruno Latour and his concept of *Dingpolitik* (2005). Developing out of the work that he undertook with Noortje Marres (2005), Latour's politics can be defined as gathering around an issue, where the focus is on concerns that move people to action rather than on power relations (Latour, 2005; Marres, 2004; Marres, 2005; Marres, 2013). While Latour does not explicitly define power, an analysis of his politics as issue formation lays bare that he perceives power as latent—as a potential that is realized when an issue generates a public around that concern. Below, I discuss Latour's political philosophy of technology and explore what politics as issue formation can offer for a non-*power over*-based politics of technology.

A Viable Alternative: Issues and *Dingpolitik*

With his notion of *Dingpolitik*, Latour aims to show that politics is about what matters, where the concept of matter refers to things, also termed as artifacts. For Latour, it is these matters (or things) that bring humans and nonhumans together: Humans are concerned with a state of affairs in the world. Latour's wide notion of matters covers all artifacts and technologies—essentially, all things that have been designed and created by humans. However, Latour denies the modern distinction between an actor (the subject) and a thing (the object) and argues that both humans and artifacts are *actants*. For Latour, there is no distinction between these two, especially when considering how scientific knowledge is produced. In a laboratory, for example, the microscope, the tubes, and the table in the laboratory have the same impact on how the knowledge is generated as the human *actant*, according to Latour (1987).

In his later writings, Latour went on to argue that politics also cannot be conceived of only as a human affair, claiming that artifacts are deeply involved in the process of politics. For Latour, a parliament should act like one collective assembly that includes artifacts as well as people. Because artifacts are not simply "things," they should be allowed to "speak" in these parliaments, and influencing the political process cannot be considered to be only a human affair. What Latour offers is a complete transformation of the basic framework of how we do politics by involving artifacts—and nature itself—as agents in the parliament. For example, if global warming is being discussed, the parliament should not only represent the people and their views and perceptions but also the phenomenon of global warming itself (Verbeek, 2020).

Latour's aim has been to give nonhuman entities an active place in society and politics. For this purpose, he has argued for a dynamic understanding of agency in which everything that *is* in the world acts in a certain way, and this changes how society is shaped. In this dynamic understanding, there is also no distinction between artifacts and human beings. For Latour, everything is mediated; there is no encounter in the world that does not involve interwoven relations between actors, including both human and nonhuman entities (Harman, 2014). Latour believes that human interactions can vanish over time. If the existence of some sort of original state of nature is accepted, the primitive engagements of that time have already vanished, since there is nothing durable in the manner of an artifact or an institution that can generate knowledge about that time today. Mediations that exist between human actors can only be durable when nonhuman actors are also involved. For example, a word can be written on a wall, on paper, or on stone, but in each case a nonhuman actor is involved (Latour, 1994). This emphasis on the importance of nonhuman entities and his conviction of the importance of mediation lead Latour to a political philosophy that is radically different from any school of thought within the field.

Dingpolitik means that politics is not only a human affair: things are actively involved in the process. Generating an issue and attracting attention to a matter are not only in the hands of the humans; according to Latour, things also possess this capacity (Latour, 2005). For Latour, each object has the capability to gather around it an assembly of relevant parties to debate and discuss, and these objects not only gather but may also offer a variety of ways of approaching the issue, reaching a solution, and forming an agreement (Latour, 2005). According to

Latour, democracy should be oriented toward objects and should detect what they bring to the fore as much as it should be concerned with detecting the relevant parties and procedures for the debate (Latour, 2005).

Latour also makes a radical claim regarding political matters with his distinction between matters of fact and matters of concern. He states that politics does not revolve around facts (things that cannot be debated or that claim to be the truth) but rather around matters of concern. It is these matters of concern that bring people together, according to Latour. He believes that humans are more connected to one another when faced with their worries, the issues they care about, and the things that concern them (Latour, 2005). He uses the verb "assembling" to define the gathering of human and nonhuman entities around a matter of concern. As Verbeek (2020) claims, in an almost Heideggerian fashion, Latour builds on the etymology of the word "thing" or "ding":

> Originally, this word meant "assembly" or "gathering," and also came to indicate the *place* of gathering, or the "parliament." At the same time, the word indicates a concern that people have—the "thing" that is at stake (in Dutch: "in het geding"). *Ding-politik*, therefore, is a conception of politics as the gathering of people around something that is at stake for them: a public and its issues.
>
> (Verbeek, 2020, p. 150)

The nonhuman entities, in this context, are not only things that people gather around or in, like a parliament building, but they are also the causes of concerns and diverse opinions. By claiming this, Latour focuses his attention back to the *things* themselves and points toward the issues they raise and bring about. According to Latour, these assemblies of human and nonhuman entities can appear anywhere, at any time, wherever the collective of human and nonhuman actors gathers. Even though there is always a physical dimension to the gathering of an assembly, it does not necessitate a designated spot. For Latour, this collective of human and nonhuman entities *is* the public, and the public creates the assembly according to the matters of concern it gathers around. Thus, with his *Dingpolitik*, Latour argues that politics is not only a matter of concern for states, experts, or bureaucrats but also for the public.

Inspired by his student Marres's (2005) reinterpretation of Dewey, Latour proposed a new angle to define the political:

> Whatever the term one wishes to use—object, thing, gathering, concern—the key move is to make all definitions of politics turn around the issues instead of having the issues enter into a ready-made political sphere to be dealt with. First

> define how things turn the public into a problem, and only then try to render more precise what is political, which procedures should be put into place, how the various assemblies can reach closure, and so on.
>
> (Latour, 2007, p. 815)

As can be seen, in Latour's terminology, an issue, the thing that people gather around, occurs when a matter of fact is translated into a matter of concern. When the life cycle of an issue is finalized, the public around it will also cease to exist. The public that is generated around one issue will dissolve, and another public will emerge around another issue, and so on. Latour's political theory of technology is about the agreement between several entities that form a public and the empowerment this agreement brings about.

Latour's politics of technology considers politics primarily as a manner of dealing with an issue, of solving a problem. One specific public that is generated around one specific issue may dissolve, ceasing to be that specific public after the issue is resolved, but the interactions and mediations that were realized within that public do not go away. Those mediations influence the disposition of the human actors and nonhuman *actants* toward the world. The same actors and *actants* may gather around a new issue, but because of the previously realized mediations, they will not be the same human actors and nonhuman *actants*. Latour pays a substantial amount of attention to the interactions between the nonhuman *actants* and human actors and how these interactions generate new entities, but still, dynamics within these entities—the struggles, identity claims, and different motives for self-realization—do not receive sufficient attention.

In Latour's politics of technology, personal and common perspectives that people create through their interactions on the issue at hand seem to be dissolving into the concept of public for the sake of achieving closure for the issue. Ultimately, the interactional aspect of Latour's politics is focused on solving a problem, and this causes Latour to inadequately thematize the technologically mediated character of human subjectivity and intersubjectivity.

A New Direction

This close examination of the prominent figures within the philosophy of technology exposes the two issues that a new political philosophy of technology has to overcome. First, such a philosophy must adopt a non-*power over*-based definition of politics in order to adequately investigate political human–technology relations. *Power over*-based approaches only expose relations of domination as relevant to politics that occur between humans and technology. The full scope of human–technology relations, and the relevance of this relation to politics, remains unexplored. As already discussed, relationality—in other words, interaction—is the basis of any power relation that can emerge. So even if one intended to understand politics only in terms of *power over*, one would still need first to grasp the basis of the power relation, which is interaction. To explore the role of technology in politics, the investigation first needs to be focused on interaction and the role of technology in this interaction. Second, this philosophy must provide a perspective that thematizes the dynamics within the public that makes up a political community, while accurately acknowledging the intricate, dynamic relationship between the subject, intersubjectivity, and the material conditions in which these interactions take place.

Hannah Arendt: A Guide for a New Political Mediation Theory

As a complement to power- and issue-driven politics of technology, this book introduces a new framework that is rooted in the ideas of Hannah Arendt and analyzes the political influence of technology through its impact on political hermeneutics and intersubjectivity. Arendt's political hermeneutics is helpful to focus on the interactional aspect of politics, thus allowing an understanding of politics that can move beyond politics as *power over*. In addition, Arendt's understanding of intersubjectivity, since it is based on a dynamic relationship between the self, the world, and other people, leaves room to investigate the role of the material conditions in this intersubjectivity. Developing this alternative account of the politics of technology based on Arendt's political theory requires a theory of technology that can concern itself with how the world becomes politically meaningful. When a set of relationships erupts between humans and technologies, such a sudden eruption means that those relationships arrive without predetermined notions that one intuitively would assign to politics.

Hannah Arendt's political theory is a rebellion against common definitions of the aim and matter of politics itself. Arendt shows that the current definition of politics—by its problem-solving and power-oriented character—fails to take its more primordial meaning into account: people forming a community. By separating politics from achieving goals and solving issues, Arendt argues for an exercise- and debate-oriented politics that is concerned with the maintenance and care of the shared, durable world (Knauer, 1980). By deconstructing the basic notions of what politics stands for, Arendt opens room for an analysis of the relationships that emerge between humans and what those relationships mean for politics. Arendt's political theory understands power as an end result of *human interaction*. In Arendt's view, power is not the ultimate defining character of political life, even though—in her specific definition—power is deeply connected to politics. According to Göhler, Arendt positions power as a purely self-referential decision (Göhler, 2009). Arendt does not consider power to be in the possession of an individual but rather to be in the possession of a group—in other words, a community of individuals. For Arendt, power is a potential that can be realized only through communication (Göhler, 2009). Göhler argues that in that sense, in Arendt's formulation, power is empowerment: It is "people gaining the ability of autonomous action" (Göhler, 2009, p. 32). According to Arendt, power "is always as we would say, a power potential, and not an unchangeable, measurable, and reliable entity like force or strength" (Arendt, 1998, p. 200).

By defining power as a *potential*, Arendt stresses a fundamental relationship between power and action. Power can only become something "actual" when people act together; it is what "springs up between men when they act together and vanishes the moment they disperse" (Arendt, 1998, p. 200). When human beings congregate and act together, even though they inevitably have different worldviews and stances on how to live or how to act—perhaps in the form of a protest or in the form of passing a bill in parliament—at that moment, Arendt believes, power is actualized.

The central category in Arendt's political theory is "interaction"; in fact, according to her theory, power should be seen as the product or manifestation of human interaction. When humans interact with one another, this always happens in a space: There is always an artificial setting, whether material or digital, physical, or nonphysical, that is the background to this human interaction.

Understanding politics in terms of *interaction* that requires a public space has important implications for the ways in which the political significance of technology can be conceptualized. However, Arendt's political theory alone is not adequate for developing this alternative account of the politics of technology since she did not really elaborate on the influence of materiality in politics.

Technological mediation theory proves to be a perfect fit for this alternative account of the politics of technology. It sees technologies as having an *in-between* role, mediating between humans and the world. By situating technologies in this way, this theory provides an analysis of the role technologies play in humans' experience of, existence in, and perception of the world. In this perspective, technologies are not only viewed as functional or instrumental but as mediators of human experiences and practices. In addition, technological mediation theory is based on a postphenomenological approach: The starting point of analysis within this theory is technologies themselves, rather than the application of philosophical theories to technologies. In this sense, technological mediation theory gives room for an analysis of a wide variety of human–technology relations that start from the relations which emerge from human–technology interactions.

Technological Mediation

Technological mediation theory can complement the missing material dimension in Arendt's political theory in order to develop an alternative account of the politics of technology, in two ways. First, as Verbeek shows, technological mediation theory argues that technologies can embody forms of power in a way that has a hermeneutic dimension (Verbeek, 2020). Technologies help shape our interpretative frameworks and are often also central to conflicts between interpretations (Verbeek, 2020). Technological mediation theory adopts a dynamic perspective on power, which allows the analysis of human–technology relations not to be limited only to relations of domination but also to focus on the generation of interpretative frameworks. Second, technological mediation theory assumes technologies are political since they organize political interactions (Verbeek, 2020). Verbeek argues that technologies enable people to come together (or prevent them from doing so) in a public space, which itself mediates the character of these interactions. According to Verbeek, it is evident that technological artifacts and systems shape human behaviors and frames of interpretation. (Verbeek, 2005). Verbeek argues that when technologies mediate our interpretative framework, they also mediate the cognitive basis from which we make decisions (Verbeek, 2020). This inevitably influences political interaction. Understanding the technologically mediated character of political interaction can help expose how technologies enable or disable political communities.

Technological mediation theory has been criticized for lacking a political dimension (Feenberg, 2009; Kaplan, 2009). This lack, according to Kaplan, stems from the theory's strong focus on individual examples, in which subjects and objects are co-constituted. Kaplan claims the theory avoids an exploration of how technical systems influence the relations not between individual actors but between society and the world. The theory of technological mediation has the potential to form the basis of a comprehensive political analysis of technology that goes beyond the prevailing power orientation. Arendt's political theory can bring the missing political dimension to technological mediation theory. In that sense, the alternative political philosophy of technology that is developed in this book synthesizes technological mediation theory and Arendt's political theory.

Further articulation of the Arendtian account of the politics of technology requires both a conceptual and an empirical investigation. Arendt, in her prologue to *The Human Condition*, writes, "What I propose therefore is very simple: it is nothing more than to think what we are doing" (Arendt, 1998, p. 5). Following Arendt's proposal, this book adopts an empirical lens at what I am doing, an explorative way of getting in touch with reality that I will later conceptually scrutinize. Even though this open exploration does not aim toward normative conclusions in itself, it sets the theme for them and allows me to make visible the framework I develop so as to investigate the political significance of technology in a more adequate way. To this end, the empirical exploration in this book focuses on a case study based on the Gezi Park protests.

Starting on May 30, 2013, Turkey's largest city, Istanbul, witnessed public unrest in its central area, Gezi Park. Before sunrise on May 31, police forces entered the park and violently dispersed the protesters who were occupying it in an attempt to stop its demolition as part of a municipal urban renewal project (Gürcan & Peker, 2015). During the protests, social media platforms—especially Facebook and Twitter—played a major role both in offering a space for the dissemination of information (nationally and internationally) and for facilitating communication among protesters (Chrona & Bee, 2017). Hundreds of thousands of people across the country took to the streets in a spontaneous and collective response to what they perceived as the government's increasing authoritarianism and conservatism (Gürcan & Peker, 2015). With this, a simple urban park became representative of the resistance to the growing authoritarian and socially interventionist Turkish government. (Gürcan & Peker, 2015). Suppressed by excessive police violence, the protesters had left the park by June 15 (Gürcan & Peker, 2015). The park and the 2013 protests became synonymous; hence, to understand how the protest's dynamics proceeded, it is important to understand what the park itself contributed to the development of the protests. In this way, the Gezi Park protests present a curious intersection between political theory and philosophy of technology that includes not just the space for digital contestation but also the spatial and cultural arrangements that coshaped the protests.

In this book, the role of empirically investigating the Gezi Park case study is explorative and supportive, giving the spirit and reality check to the philosophical analysis. To this end, the empirical exploration that I undertake builds on the existing studies analyzing the events of the Gezi Park protests (Baban & Güzel, 2015; Chrona & Bee, 2017; Eken, 2014; Haciyakupoglu & Zhang, 2015; Kuymulu, 2013; Tufekci, 2013; Tufekci, 2017; Turekci, 2015; Özbank, 2013). Occasionally, I also draw on semistructured interviews I conducted myself with the citizens of Istanbul and participants of the Gezi Park protests to illustrate and qualify the conceptual analysis, making it more lived and real. The version of empirical philosophy I conduct here is therefore relatively "lightweight" in that it does not use empirical data as a groundwork for conceptual elaboration. Rather, it uses elements of ethnography—through conversations with the protest participants—for illustrative purposes and to set the stage for the analysis.

Regardless of the modest scope of my empirical contribution, it is essential to unpack the role of the public park in the creation of a heterogeneous political community and in the creation of spaces for contestation. In doing this work of unpacking, I follow Annemarie Mol (2021), who suggests that philosophers engage with the real practices in which objects are involved—objects that philosophy otherwise addresses from a distance. Presenting the case study analysis and illustrative quotes from the interviews in this book thus helps me to position the philosophical concepts in real life and challenge them from within the practices surrounding them.

Outline

This book consists of two parts. The first part is composed of Chapter 1 and Chapter 2 in which I lay the foundations for this alternative, Arendtian account of the political philosophy of technology. In the second part of the book, I develop the concept of technopolitical mediation, synthesizing technological mediation theory and Arendt's political theory. I demonstrate that technopolitical mediation occurs in two steps: technological mediation of common sense, which I explore in Chapter 3, and technological mediation of intersubjectivity, which I explore in Chapter 4. Through the analysis of technological mediation of common sense and technological mediation of intersubjectivity, this book shows that technology plays a crucial role in enabling the authentic development of values, both on an individual level and within a broader community.

Thus, Chapter 1, "Politics as Interaction: Hannah Arendt and Political Theory," introduces important components of Arendt's political theory and clarifies how these concepts are related to one another. According to Arendt, politically active life is *vita activa* (Arendt, 1998). This *vita activa* consists of three different modes of being human: *animal laborans, homo faber,* and *zoon politikon* (Arendt, 1998). These modes of being are also connected to forms of activities: *animal laborans* works and takes care of humans' biological needs; *homo faber* creates durable things, such as buildings that provide shelter for humans; and *zoon politikon* acts, appearing in public as a distinct, unique being via engaging in discussions with fellow human beings in public spaces. The action of *zoon politikon* is what reveals the world as politically meaningful for Arendt; thus, politics can only occur when *zoon politikon* acts and interacts.

Arendt interprets politics as *praxis*—that is, an activity that does not need an external justification for its existence: It is an end in itself. Through its appearance in the world, *praxis* is accompanied by human plurality and freedom. These concepts are, on the one hand, what secures the possibility of *praxis*'s appearance; on the other hand, they are what *praxis* aspires to achieve in the condition of their nonexistence. By establishing *praxis* as politics, Arendt argues against conceptualizing politics as *poiesis*—that is, understanding politics not as an end in itself but as instrumental to achieving an aim or result. Building on an extensive exegesis of Arendt's work, the chapter concludes by arguing that today, politics is being considered as if it were a form of technology. While arguing this assumption to be wrong, the chapter concludes that technologies can be political since they condition the way in which politics is carried out.

In Chapter 2, "Arendt, Phenomenology, and Technology," I argue that *poiesis* is as essential as *praxis* for politics. By maintaining the distinction between *poiesis* and *praxis*, I show that they are in fact intertwined rather than opposed. I show that a close inspection of the meaning of the concepts *praxis* and *poiesis* for Arendt necessitates a close inspection of her method of analysis, which is phenomenology, specifically hermeneutic phenomenology. In fact, Arendt's political theory is a hermeneutic phenomenology of the political (Borren, 2013). I demonstrate that "Arendt's hermeneutic phenomenology" (Borren, 2013) allows her to situate the world as the basis of human understanding and meaning making. This observation leads to grasping Arendt's notions of the common world and common sense as the building blocks of her political theory. The fact that Arendt positions common sense as both "world building" and "world disclosing" (Borren, 2013), I demonstrate, suggests a closer relationship between *poiesis* and *praxis*. This leads me to argue that it is *poiesis* that creates the world on which human life depends. I argue that the world Arendt positions as essential for human life is indeed artificial but also material. Following this, I show that Arendt's insistence on the opposition of *praxis* and *poiesis* stems from her observations regarding modern technology and mass society. These observations lead her to position modern technology as antithetical, if not irrelevant, to politics. However, following Arendt's understanding of the cultural products of *homo faber*, I argue that she can guide us to formulate an interaction-based politics of technology. Following this, I show that the cultural products of *homo faber* not only stabilize the appearances of *praxis* but also inspire further appearances of *praxis* because they are an embodiment of the meaning that the world has been given at their time of creation.

In Chapter 3, "Technological Mediation of Common Sense," I investigate how technologies mediate our common and political interpretative frameworks. The chapter starts by arguing that postphenomenology, rather than phenomenology, can help Arendt's political theory to expand the implications of the influence of *poiesis* on *praxis*. This is because it is postphenomenology that investigates the hermeneutic implications of materiality. I show, through the postphenomenology of technology, that things (artifacts, objects, and technologies) have a hermeneutic dimension; because of that quality, they mediate human experience and how humans have access to their world—in other words, how humans make sense of their world. Following on this insight, I refer to Olya Kudina's (2019) scheme of the hermeneutic lemniscate to explore the mediation of common sense by technologies. I demonstrate that Kudina's hermeneutic lemniscate presents a new principle of human interpretation and sense-making, namely that human interpretation and sense-making are technologically mediated. Therefore, the model helps in explaining the role of the cultural and moral history of the person engaging with the technology in question. Subsequently, I analyze how *poiesis* mediates *praxis* by providing scripts from the interviews that I conducted with the participants of the Gezi Park protests. Through these extracts, I show how the material constructs of *homo faber* in the form of technologies have the same hermeneutic power that cultural products of *homo faber* have. Both of them contribute to our common sense and mediate the human experience of the world. I end the chapter with a synthesis of the political hermeneutics of Arendt and the material hermeneutics of technological mediation theory.

In Chapter 4, "Technological Mediation of Intersubjectivity," I discuss the mediating role of technology in intersubjectivity. I show that the manifestation of the authentic and free self is either fostered or suppressed due to the intersubjective relationships within the community the individual is engaged in. I argue that if this community has limited to no possibilities for the self to take personal responsibility and express its unique perspective on matters that concern the individual, it will suppress the possibility of the manifestation of the authentic and free self. Building on the distinctions presented, through an analysis of the interviews I conducted with the participants of the Gezi Park protests, I focus on the role of technology in these intersubjective relationships. I illustrate that there is an intrinsic relationship between the authenticity of the community and the material space that hosts this community. Finally, I provide an additional aspect of community that is missing in Kudina's (2019) hermeneutic lemniscate. I show that the material space where people meet mediates both the way in which a community is established and the way in which an individual becomes a part of a community.

The book concludes by arguing that technologies do much more politically than simply exerting power over individuals: They also condition, frame, instigate, and organize politics. Technologies inform the way in which political communities are formed by mediating our political interpretations and interactions. Technologies mediate common sense—in other words, the common interpretative frameworks of humans—which in turn is reflected in the common interpretative framework of the political community. Technologies also mediate intersubjectivity. An individual's political decisions, and how they express them, are either fostered or suppressed due to the intersubjective relationships within the community in which the individual is engaged. There is an intrinsic relationship between the political structure of the community and the material space that hosts this community. This material environment conditions and organizes the structure of the political community in which the individual is engaged. In this way, the way in which people express themselves politically not only depends on the cultural, institutional, or intersubjective conditions but also on the material conditions.

Part One

1

Politics as Interaction: Hannah Arendt and Political Theory

A limited conception of politics that is informed only by a notion of *power over* misses a grasp of the intricate relationship between humans and technology—and what that would mean for politics. In order to go beyond this limited approach, there is a need for Hannah Arendt's interaction-based political theory. However, this necessitates a thorough examination of politics as interaction in the first place.

This chapter, therefore, sets out the foundations of Arendt's political theory as a prelude to building a new approach within the political philosophy of technology. It interrogates Arendt's interpretation of the Aristotelian concepts of *praxis* and *poiesis* and their relationship to politics. Based on this discussion, the chapter shows that traditionally, politics has been coupled with an instrumental, goal-oriented approach toward the world that is inherent in *poiesis*, our human capacity that is guided by *techne*. Following Arendt (1998), I show that this coupling limits the realm of politics and causes a misinterpretation of the definitions of features that make up the political realm, such as plurality, freedom, action, and power. Finally, I show that once the basis of politics is conceptualized as *praxis*, this enables a political space, a *polis*. This *polis* enforces the citizen individual acting in the world as a free agent connected to their community through shared concerns, experiencing freedom among a plurality of perspectives through interaction.

To lay the foundations for this argument, the first section of the chapter delves into Arendt's criticism of Plato's influence on conceptualizing politics as *poiesis*—that is, understanding politics not as an end in itself but as instrumental to achieving an aim or result. Following Arendt, I argue that politics was based on *praxis* before Plato's intervention, and I call for revitalizing politics as *praxis*.

Having established that Arendt's political theory is based on politics as *praxis*, the second section of the chapter starts with introducing important components of this political theory and clarifying how these concepts are related to one another. According to Arendt, a politically active life is *vita activa* (Arendt, 1998). This *vita activa* consists of three different modes of being human: *animal laborans, homo faber*, and *zoon politikon* (Arendt, 1998). These modes of being are also connected to forms of activities: *animal laborans* labors and takes care of humans' biological needs; *homo faber* works, creating durable things such as buildings that provide shelter for humans; and *zoon politikon* acts, appearing in public as a distinct, unique being via engaging in discussions with fellow human beings in public spaces. The action of *zoon politikon* is what reveals the world as politically meaningful for Arendt; thus, politics can only occur when *zoon politikon* acts and interacts. Because Arendt interprets politics as *praxis*, she sees it as an activity that does not need an external justification for its existence: it is an end in itself.

After demonstrating these main distinctions between the activities that make up *vita activa*, in the third section, I explain how this form of life can be realized. This discussion reveals that plurality, freedom, action, and power are essential features of a politically flourishing life. I show that these concepts are intimately related to one another, and that together they are all essential in the appearance of political activity as *praxis*. That is, they are essential for the practical manifestation of how the idea of the political appears to one and how one puts that idea into practice. These concepts are, on the one hand, what secures the possibility of *praxis*'s appearance in the world; on the other hand, they are what *praxis* aspires to achieve in the condition of their nonexistence.

The chapter concludes by arguing that conceptualizing politics as interaction allows room for a thorough exploration of the influence of technology in politics. I argue that only through maintaining the distinction between *poiesis* (that has a result manifested as technology) and *praxis* (that has a result manifested as politics) can we investigate the intricate relationship between the two.

Plato's Transformation of the Polis

In *The Human Condition* (Arendt, 1998), Arendt begins to unravel the meaning of politics, tracing its beginning to Plato, whom she criticizes because she alleges that he replaced the Greek concept of *praxis*[1] (action) as the basis of politics with *poiesis* (fabrication). Plato's motive for this replacement, according to Arendt, was to announce the life of contemplation—*bios theoretikos* in Greek—as the mode of living that is more worth pursuing[2] than *bios politikos* (life of action). Arendt, following the distinction between *bios politikos* and *bios theoretikos*,[3] invents her own two fundamental categories where human life unfolds: *vita activa* and *vita contemplativa.* While *vita contemplativa* is the withdrawal from the world, the public, and political affairs, *vita activa* is the life devoted to them (Arendt, 1990).

According to Taminiaux (1997), Plato not only presents *vita contemplativa* as superior to *vita activa* but also situates *vita activa* as a form of life that is experienced to "assist" (Taminiaux, 1997, p. 24) *vita contemplativa.* Plato presents the philosopher, the one who devotes his life to contemplation, as the one who also needs to carry out physical activities to remain alive. These activities, such as maintaining the body, providing oneself with shelter, and maintaining the silence and calm of contemplative life, necessitate the philosopher's having a political life that is organized and peaceful (Taminiaux, 1997). For Plato, however, *praxis* is in contradiction with providing the calm and peace the philosopher needs because of the qualities that are inherent to it.

Plato, according to Arendt, especially after the death of Socrates, wanted to strip action (*praxis*) of its political character because of its inherent qualities and searched for ways to construct a new form of political activity (Arendt, 1998). To that end, Plato first blurred the distinction within the activities of *vita activa,* namely *poiesis* and *praxis,* so that he could situate *vita contemplativa* as the superior mode of living (Taminiaux, 1997). Arendt here follows an etymological path to reach an understanding of how *praxis* was subjugated to *poiesis.* First, she goes back to the Greek and Latin roots of the verb "to act":

> In Greek the verbs *archein* ("to begin," "to lead," finally "to rule") and *prattein* ("to pass through," "to achieve," "to finish") correspond to the two Latin verbs *agere* ("to set into motion," "to lead") and *gerere* (whose original meaning is "to bear").[4]
>
> (Arendt, 1998, p. 189)

Arendt says that *praxis,* because of the Greek and Latin verbs that it contains, seems as if it is divided into two parts: to begin, and to finish what has been begun (Arendt, 1998). For Arendt, these verbs are interrelated, but both the Latin *gerere* and Greek *prattein* assumed the sole definition of action (Arendt, 1998). *Archein* was diminished to only one meaning, namely to rule, and *agere* only meant to lead, being stripped of its meaning of setting something into motion (Arendt, 1998).

According to Arendt, Plato opened the way for a new form of relationship to be the defining character of politics by differentiating two interconnected modes of action: *archein* (beginning) and *prattein* (achieving)—ruling and being ruled (Arendt, 1998). For Arendt, Plato wanted to make sure that the one who carries out *archein* would remain in control of what they had started alone, without the help of their fellow human beings (Arendt, 1998). *Praxis* in its original sense depended on both *archein* and *prattein*, requiring the existence of others for its execution (Arendt, 1998). When action appears, as noted above, it establishes a set of relationships. These relationships appear in the form of a reaction to the action. Therefore, for action to be completed, it is essential that there be an audience that reacts to the act that has been performed. For example, a famous speech such as Martin Luther King Jr.'s "I Have a Dream" speech required an audience. If there had been no audience to appeal to, King Jr. would not have been able to reveal the world as it appeared to him and reveal himself.

By isolating *archein* as an activity in itself, Plato managed to constitute beginning and acting as separate activities and thus established the beginner as the ruler who "does not have to act at all (*prattein*), but rules (*archein*) over those who are capable of execution" (Arendt, 1998, pp. 222–3). According to Arendt, Plato's proposal, in light of this search for a new political form of life, was the philosopher-king, who was solely responsible for attending to public affairs. By situating the philosopher as the king, Plato banished the citizens of the *polis* from attending to the common affairs with their fellow human beings (Arendt, 1998). According to Arendt, this relationship of ruling and being ruled had existed in the privacy of the household for the ancient Greeks, but experiencing this form of relationship in the *polis* was foreign to them.

Plato's division of *archein* from *prattein* meant the division between thought and action, which caused the transformation of the *polis*. Since then, the *polis* has been constructed via the relationship between ruling and being ruled, where action cannot play any part as it did in the original form of the *polis* (Arendt, 1998).

Plato's substitution of rulership for action became stronger when he interpreted *praxis* (action) as *poiesis* (fabrication). *Praxis* had already lost its validity and meaningfulness when it was separated from thought, but Plato applied his doctrine of ideas to the political realm as well, where ideas were constructed as standards. Before his interest in politics, Plato defined ideas as variations of the beautiful, as "what shines forth most (*ekphanestaton*)" (Arendt, 1998, p. 225). However, later in the *Republic*, Arendt argues, Plato transformed his concept of ideas to variations of "the good," and he then situated this concept of the good as the highest single idea of the philosopher (Arendt, 1998, pp. 225–6). Just like the objective standards that guide the craftsman toward his work when he is in the process of making, the good becomes the objective standard of the philosopher-king when he is ruling. For Arendt, this "good" is not the highest ideal sought by the philosopher in the cave via contemplation, but rather the highest idea attained by the philosopher-king who seeks to turn it into "objective" certainties (Arendt, 1998, p. 226). By situating the good as the highest objective standard for ruling, Plato aimed at eliminating the private or personal interest of the rulers, which could run counter to the communal interest of the *polis* (Arendt, 1998). Plato, by separating *archein* and *prattein*, first eroded the meaning and validity of *praxis* and later completely removed it from the realm of political human affairs by substituting it with *poiesis*.

Dossa (1989) argues that Plato's demand to contain action, and his replacement of it with *poiesis*, resulted in a complete change in understanding of life: life in the polis became subordinate to life in contemplation (Dossa, 1989). The conception of politics in the image of *poiesis* (which is guided by *techne*) caused political life to be conceived in terms of "in order to" relations, and political life began to validate its existence in order to serve a higher end, namely the life of contemplation. This, according to Arendt, became the basic frame of reference for all Western traditional political thought (Arendt, 1998). Because of this substitution, politics itself started being understood in terms of "in order to" relations (Arendt, 1998). The citizens need not bother themselves with appearing in the public realm, since only the ruler is allowed to attend to public affairs (Arendt, 1998). The ruler, however, exercises his rule only to provide "stability, security and productivity" to his citizens (Arendt, 1998, p. 222).

For Arendt, as long as we humans believe that politics is an exercise of means–ends relationships, humans will be unable to prevent anyone from using all means that are available to them to reach the ends they pursue (Arendt, 1998). Arendt aims for a complete break with this conception of politics, which replaced the Greek notion of *praxis* (action) with *poiesis* (fabrication). Her political theory offers a new beginning via a recovery of the basic notions that political theory is concerned with, such as political action, power, and the public.

Vita Activa

In order to recover these basic notions that political theory should be concerned with, Arendt returns to her fundamental understanding of the main concern of politics: the human. Arendt perceives human existence as a conditioned existence, which means that humans are always influenced by certain conditions that change and shape the way they exist in the world. The moment a person interacts with something or someone, they enter into a relationship; they influence one another, and thus the other assumes a role in the way in which one exists (Arendt, 1998). The conditions that influence humans can be either natural or artificial, and they can change, but the fact that humans are always influenced by these conditions does not change (Arendt, 1998).

Arendt uses the term "conditionality" to designate activities that actualize conditional structures by generating specific experiences and specific forms of meaning (Loidolt, 2018). Arendt's understanding of conditioned human existence does not depict a world where individuals draw agency from "fixed entities, structures or meanings" (Loidolt, 2018, p. 111). Instead, conditionality refers to a transcendental relationship between the subject, the object, and the activity that is being carried out (Braun, 1994, as cited in Loidolt, 2018). This shows that, for Arendt, humans are both influenced by and shaped by the actions they undertake and the material environment that surrounds them.

The world we live in, according to Arendt, is not the natural earth, but rather it is created by humans and is an accumulation of things and artifacts that we ourselves create. In other words, the world as it is *exists* because we humans created it (Arendt, 1998). This world that we have created is also public. Humans do not live in the world as singular individuals; they inhabit the world together with other humans. Thus, there are two characteristics of this world that we have created: first, it is artificial, an end product of human endeavor; and second, it is public, a space that is shared with fellow human beings. Arendt argues that even though, as humans, we are conditioned beings, who we are and what we are, namely what kind of person each individual human being is, cannot be explained by referring to the conditions that influence us (Arendt, 1998). However, Loidolt claims that these conditions create an identity in the actualization of each condition and corresponding activity (Loidolt, 2018).

According to Loidolt, Arendt's concept of conditionality refers to the way in which human life unfolds within the condition and the corresponding activity (Loidolt, 2018). When humans are in a certain mode of being, the condition in this case, everything they have a relationship with, the way they are, and the way they see and perceive the world becomes meaningful within the set of relationships in that condition.

According to Arendt, this conditioned-ness is a matter of our humanness; it is not an absolute that defines us, but it is a part of us (Arendt, 1998). Loidolt explains this as part of forming an identity (Loidolt, 2018).

The conditions that influence humans, according to Arendt, are life itself, natality and mortality, worldliness, plurality, and earth (Arendt, 1998). These conditions refer to three fundamental human activities—labor, work, and action—that, as mentioned above, all take place within what Arendt calls *vita activa* (Arendt, 1998). The Arendtian concept of *vita activa* consists of three activities that are different in their nature: labor, work, and action. The three activities are inherent in every human being; in other words, every human life has the capacity to exercise labor, work, and action (Arendt, 1998). These activities also have three different conditions to which they are respectively connected: labor is connected to the human condition of life itself, work is connected to the human condition of worldliness, and action is connected to the human condition of plurality (Arendt, 1998). These three conditions, Arendt says, are also connected to two natural conditions: natality and mortality. Natality is the condition of birth—being born into this world as a newcomer—and mortality is the condition of death, the end of life, which is a universal condition for every human being (Arendt, 1998).

According to Parekh (1981), among the three activities that constitute *vita activa*, there is a hierarchy determined by "their degree of contribution to human existence" (Parekh, 1981, p. 103). This hierarchy stems from Arendt's understanding of a transcendental relationship between subject, object, and the activity that is being carried out (Loidolt, 2018). Loidolt explains that Arendt adopts this transcendental approach from Heidegger's "existentialia"[5] (Loidolt, 2018, p. 52).

According to Heidegger, "transcendental knowledge does not investigate the being itself, but rather the possibility of the preliminary understanding of Being, i.e., at one and the same time: the constitution of the Being of the being" (Heidegger, 1997, p. 10). This "possibility of the preliminary understanding of Being" (Heidegger, 1997, p. 10) or "the constitution of the Being of the being" (Heidegger, 1997, p. 10) can be translated as authenticity (Villa, 1995). This means that the hierarchy within *vita activa* is developed by Arendt in relation to the preliminary understanding of being—that is, authenticity (Villa, 1995).

The hierarchy describes the activities in relation to their conditions of possibility for authentic human existence. According to Loidolt, with *vita activa* and the activities that constitute it, Arendt develops a new phenomenological approach (Loidolt, 2018). This approach allows Arendt to locate Heidegger's "existentialia" in "worldly, bodily and intersubjective conditions" (Loidolt, 2018, p. 110). These conditions, as mentioned before, do not define us in their entirety, but they can be understood as modes of existence. These modes of existence influence the way in which the world becomes meaningful to us, since we interpret the world within the mode of existence that we are in. In everything that we do, different dimensions of *vita activa* are highlighted, and one dimension might be more dominant than the others.

Labor

Labor, the activity that corresponds to the condition of life itself, assumes the lowest position within the hierarchy of *vita activa*, since it is connected to the bodily condition of life itself. According to Arendt, labor is the manifestation of the basic behavior imposed on humans, namely the necessity to survive. By laboring, human beings satisfy their needs for survival. The activities that come together with labor serve only to sustain life (Arendt, 1998). The activity of labor has neither beginning nor end; like nature itself, it has a cyclical movement, and it is controlled by biological needs. For Arendt, to labor is therefore to be enslaved by necessity, and it is not subject to choice or decision (Arendt, 1998). This is the form of activity that humans share with all animals; thus, when laboring, Arendt claims humans are in the mode of *animal laborans* (Arendt, 1998). When we are in the mode of existence of *animal laborans*, Arendt claims, we are not our real selves because we are driven by the necessity that our physical bodies impose on us. We are not distinct human beings, since the activity we carry out has no authenticity (Arendt, 1998). This natural existence of ours has a cyclical movement. For example, satisfying hunger will never come to an end; the moment a person digests the food they eat, they will start feeling hungry again and will try to find ways to satisfy hunger, and this will continue until the person's life comes to an end. For Arendt, one cannot survive only with the activity of labor, since *animal laborans* always needs defense and protection from the natural processes. To achieve this, *animal laborans* has to take things out of nature's hands and consume them (Arendt, 1998). Humanity then cannot survive only with the condition of life itself; it also needs the condition of worldliness.

Work

The activity corresponding to the human condition of worldliness is work (Arendt, 1998). This human condition assumes the second place within the hierarchy of *vita activa*. Worldliness designates the unnaturalness of human life. The activity within this condition, namely work, interferes with nature and creates something artificial out of it (Arendt, 1998). For example, chopping down a tree in the forest and building a house with the wood is the work that we carry out. The things that we produce in this mode of being come together and make up the artificial world that we live in (Arendt, 1998). The "work" in Arendt's terminology refers to the previously mentioned term derived from Aristotle: *poiesis*.

In Aristotle's terminology, *poiesis* refers to activities that cover economic life, have predictable ends, and require practical means to be achieved (Carr, 2006). *Poiesis* is guided by *techne*, which would be translated into instrumental, means–ends reasoning in our contemporary society (Carr, 2006). This instrumental activity requires a certain type of knowledge, mastery, methods, and skills to be carried out (Carr, 2006). For example, one has to know what kind of wood to use to make a specific kind of table and for a specific purpose. Because of these requirements for *poiesis* to be carried out, this activity necessitates procedures, a blueprint, so to say, to be followed, which guides the person who carries out the activity. With such a guide, it is possible to establish the most efficient way to achieve the predetermined goal (Carr, 2006).

Poiesis is instrumental to the aim that it is directed toward. When a person takes a piece of wood with the aim in their mind of making a table, the existence of that piece of wood becomes instrumental to achieving the goal of the table. For Aristotle, in the process of achieving the table, even the creator of the table (the carpenter) becomes instrumental (Balaban, 1990). The only thing that justifies the existence of the set of relationships that surround the creation of a wooden table is the actualization of the wooden table (Balaban, 1990). According to Aristotle, argues Balaban, there is no difference between the producer and the means of production (Balaban, 1990): The one who hammers the nail is not different from the hammer, and the one who hammers also becomes a means of production. In addition, Balaban argues that it would be absurd to assume that *poiesis* is a teleological activity, since once the end result of the activity is achieved, *poiesis* ceases to exist (Balaban, 1990). This activity in this sense is "inherently contaminating," since it subordinates the activity that is being carried out to its ultimate end (Martin, 2017, p. 51).

Arendt defines the activity of work as fabrication, and as already mentioned, this activity is *poiesis* that is guided by *techne* in Aristotle's terminology. When humans are carrying out *poiesis*, according to Arendt, their mode of existence becomes that of the craftsman, *homo faber*, and they "fabricate the sheer unending variety of things whose sum total constitutes the human artifice" (Arendt, 1998, p. 136). This human artifice is objective in nature: As already indicated, even the creator of the artifice becomes an object in the process of its creation.

The fact that *poiesis* has a specific moment of beginning and a designated end is the feature that most clearly distinguishes it from the other activities within *vita activa*, according to Arendt (Arendt, 1998). Since the activity follows certain rules and a blueprint, the end result can be foreseen. This feature makes work a reliable activity. Besides, Arendt adds, the end result of work can be destroyed, and this means that the process is reversible: If one does not like the end product, it can be demolished completely. Even while carrying out the activity, one can simply stop and not produce anything. *Homo faber*, then, creates durable, tangible things that can be destroyed but are meant to remain, even when the creator of the things leaves the world. For Arendt, this durable character of the things of the world gives them their "relative independence from men who produced and used them, their 'objectivity' makes them withstand, 'stand against' and endure" (Arendt, 1998, p. 137). The durability and stability of human artifice create a break from the circular nature of life; they stabilize human life. I elaborate on the value of the human condition of worldliness and its activity, work, in more detail in the next chapter. For now, it is important to understand that this condition assumes the second position within the hierarchy of *vita activa*.

Action

The highest activity within *vita activa* is action (Arendt, 1998). This activity corresponds to the human condition of human plurality (Arendt, 1998). With action (*praxis*), Arendt points to the communal character of human life. When humans carry out *praxis*, they are *zoon politikon* for Arendt (Arendt, 1998). Work and labor are activities that can be realized in isolation, but action requires the existence of others and cannot be realized in solitude (Arendt, 1998). Arendt defines action as "to take an initiative, to begin …, to set something into motion" (Arendt, 1998, p. 177). Arendt refers to Aristotle's *bios politikos* in attributing the highest rank within *vita activa* to action. This *bios politikos* is the life that is experienced in *koinon*, meaning with the community, via *praxis* (action) and *lexis* (speech) (Arendt, 1998). In Arendt's terminology, action corresponds to Aristotle's concept of *praxis*. Taminiaux (1987) argues that *praxis*, in its original Greek meaning, was

> an activity which neither relates a living being to the eternal recurrence of life nor relates an expert to natural data which are to be worked upon or elaborated or used but which relates individuals to other individuals in the sharing of common care for a common realm, and which discloses those individuals as who they are in relation to other individuals likewise disclosed.
>
> (Taminiaux, 1987, p. 138)

Praxis in this sense is a form of doing, and its realization, the promotion of *good life*, "can only be realized in and through *praxis* itself" (Carr, 2006, p. 426). Unlike *poiesis*, one cannot theoretically specify the end result of *praxis* (Carr, 2006). The knowledge of the end result of *praxis* appears during the act itself (Carr, 2006). In that sense, *praxis* is very much context dependent; the knowledge of what *praxis* could lead to is based on an understanding of the context and how "this knowledge is being interpreted and applied" (Carr, 2006, p. 426). In sum, for Aristotle, *praxis* is

> a practical manifestation of how the idea of the good is being understood, just as knowledge of the good is nothing other than an abstract way of specifying the mode of human conduct through which this idea is given practical expression.
>
> (Carr, 2006, p. 426)

This definition also conveys that the knowledge of what good is and the knowledge of how one applies it are supportive of each other in the process of practical reasoning (Carr, 2006). This reasoning that accompanies *praxis* is *phronesis*, according to Aristotle (Carr, 2006). Unlike *techne*, which accompanies *poiesis*, *phronesis* cannot be learned in isolation from practice, and it is not a deductive form of reasoning giving a blueprint for *praxis* to follow (Carr, 2006). *Phronesis*

> can only be acquired by practitioners who, in seeking to achieve the standards of excellence inherent in their practice, develop the capacity to make wise and prudent judgements about what, in a particular situation, would constitute an appropriate expression of the good.
>
> (Carr, 2006, p. 426)

In that sense, *phronesis* "is a moral and intellectual virtue that is inseparable from practice" (Carr, 2006, p. 426). So, keeping in mind the Aristotelian notion of *praxis* and its accompanying *phronesis*, Arendt adds to this concept the manifestation of how the idea of the political appears to one and how one carries out that idea. Arendt understands action not as an end, but as the beginning of something new (Martin, 2017). Arendtian action is not about the results it may produce; it is about the act itself (Martin, 2017). This does not mean that action is not oriented toward accomplishing things; however, what action can achieve is highly dependent on the reactions of others, and therefore the consequentialism that work has, for example, is foreign to action (Martin, 2017). Action, in this sense, transcends the ends it may produce, and its uniqueness is its key quality (Martin, 2017). I elaborate on the value of the human condition of plurality and its activity, action, in more detail in the next section. For now, it is important to understand that this condition and its accompanying activity assume the highest position within the hierarchy of *vita activa*.

Until now, I have explained the activities and the corresponding modes of being that constitute *vita activa* for Arendt. For her, *vita activa* is the form of life devoted to the realm of human affairs: in other words, politics. Arendt's political actor lives the life of *vita activa* and is concerned with the world that we share with others (Arendt, 1998). In the following section, I explain the features of politics through which *vita activa* can be realized. These features are, on the one hand, what secures the possibility of *praxis*'s appearance in the world as a political reality, and, on the other hand, they are what *praxis* aspires to achieve in the condition of their nonexistence.

Features of Politics

By establishing *praxis* as the basis of politics, Arendt revitalizes the public quality of politics, situating human existence with an ontological connection to the world that can only be meaningful when the world is revealed in the manner of politics. To strengthen this basis, Arendt portrays *praxis* as strongly associated with human plurality, freedom, natality, and power—none of which has a connection with instrumentality. It is the existence of plurality, freedom (which is intimately connected with natality), action, and power that fosters politics as interaction. Below, I elaborate on these features.

Plurality

Plurality is a prior condition for *praxis* to appear, but it is also a drive for *praxis*. If the realm in which *praxis* appears does not include plurality, *praxis* aspires to realize it. According to Arendt, human plurality signifies the fact that "men, not Man, live on earth and inhabit the world" (Arendt, 1998, p. 17). The existence of plurality depends on a certain feeling of togetherness, a commonality that is shared with fellow humans in the community. This community acknowledges that every human is a distinct, authentic being, and it respects that every human being has the freedom and autonomy to appear in the world as such. In our contemporary societies, human distinctiveness seems to be based on independence and individualism. However, for Arendt, almost the opposite is true. Human distinctiveness for her is based on the feeling of togetherness with one's fellow human beings while each individual remains an authentic existence. Wherever plurality is present, it enables a public space to come into being by allowing people to meet and gather in order to perform speech and action. Plurality, in this sense, is the manifestation of distinct human beings coming together and being exposed to each other. While plurality is an ideal condition in society, sometimes it cannot be experienced. However, if plurality does not exist, this does not hinder the possibility of action to appear.

In "Totalitarian Imperialism: Reflections on the Hungarian Revolution" (Arendt, 1958), Arendt celebrates the spontaneous, unexpected action of the young Hungarians against the totalitarian regime of the Soviet Union. She claims that this revolution had no leaders and no organization, and it was not directed by anyone or anything. "The will for freedom," she writes, "was the moving force in every action" (Arendt, 1958, p. 8). So ideally, plurality is the default position in society, but when uniformity is the default position, action (*praxis*) strives for the realization of plurality. So, on the one hand, plurality is presupposed for action to appear; on the other hand, it is something that action continuously strives to realize in the condition of its nonexistence. Below, I explore different aspects of plurality as it was conceptualized by Arendt.

a) Plurality's Equality and Distinctness

For Arendt, the characteristic features of human plurality are equality and distinction. With action, humans reveal their distinctness and uniqueness, their authentic existence in the world (Arendt, 1998). The sum total of these unique, distinct beings creates plurality. The other characteristic of plurality is equality, which Arendt takes up from the ancient Greek understanding. Equality, or *isonomie*, according to Arendt, is the "very essence of freedom: to be free meant to be free from the inequality present in rulership and to move in a sphere where neither rule nor being ruled existed" (Arendt, 1998, p. 33). Arendt argues that in the ancient city-state, the public realm was reserved for individuality, in the sense that each citizen of the city strove to show himself, to rise above the others (Arendt, 1998). This was only possible when one was allowed to live among one's "equal" peers. In the *polis*, the ancient city-state, this fiercely agonal spirit was spread throughout every human being. They all wanted to show their unique, distinct beings *par excellence* to their peers, their equal fellow human beings (Arendt, 1998). Arendt perceives equality as an expression of freedom in the sense of being free from being ruled and having the capacity to rule one's own self.

Human plurality in Arendt's political theory, then, is the recognition that humans do not live in singularity; one must acknowledge that one lives in a world that is shared with other singular beings. The existence and the survival of the world that we share in common not only depend on how an individual acts in this world but also on how that individual acts together with others. Every human has their own ideas, they perceive things from their own perspective, they are all able to imagine a future for themselves, and they all have the capacity to act (Arendt, 1998; Canovan, 1983).

Arendt was not blind to the fact that recognizing and allowing every human being to show their distinctness in the public or political realm carries within it the possibility of anarchy and destruction.[6] However, there is the possibility of establishing institutions to ensure that every human can show themself via reaching mutual consent and agreement (Canovan, 1983). Instead of reducing this consent and mutual agreement to one single will, Arendt argues that diverse, unique individuals can still be united and create a common world (Canovan, 1983). According to Canovan, Arendt here stresses the fact that every human has their own opinion and their own judgment, derived from their own experience and perception of the world, and these can be directed toward consent and agreement when they are coupled with political compromise for the sake of the shared common world (Canovan, 1983). However, Arendt refuses the conclusion that, when this public deliberation and discussion occur, and where reason and rationality rule, the citizens will inevitably reach a single truth (Canovan, 1983). For Arendt, expecting the citizens to reach a single common truth would be against the spirit of human plurality. Arendt's point is that these individuals can compromise and act out of a sense of loyalty for the sake of the shared common institutions that guarantee their freedom and equality (Canovan, 1983). They can be a community that is fully united, not as a single entity in the sense that they think alike, but because they recognize the fact that they share the same public or political realm, respect the rules and regulations that people have agreed upon, and are committed to accepting compromise for the maintenance of this shared common world (Canovan, 1983). Plurality, for Arendt, is not only a feature of politics that guarantees equality and distinctness, but it also guarantees that humans have access to reality. Below, I unpack the intricate relationship between plurality and reality.

b) Plurality as Reality Check

For Arendt, plurality does not necessarily mean that one can say or do whatever one wants in the realm that is shared with others. To a certain extent, of course, every human being is allowed to share who they are and how the world appears to them. But the aim of plurality is to turn the subjective perspectives of how the world appears to one into an agreed reality. In one's own experience, one only knows how the world appears to oneself. To grasp how the world really is, one must be able to see and hear how the world appears to others too. Different people with different perspectives about the world can form different opinions. However, the validity of these opinions depends on free agreement and consent. Thus, if we want to perceive the world in its totality, if we want to "experience the world as it '*really*' is" (Arendt & Kohn, 2005b, pp. 128–9), we can only do so by experiencing it as something that lies between us and our fellow human beings, and as something that separates us from our fellow human beings.

The world, in its objectivity, shows itself differently to each human being. It becomes something meaningful and comprehensible as long as various people can express their differing perspectives and experiences with each other (Arendt & Kohn, 2005b). Human plurality in that sense allows humans to check reality constantly. What is real is not something that one can perceive within one's own solitude; what is real can only be experienced with others, via checking if we all see the same thing at the same time. Seeing things from the perspective of others allows us to examine our position continuously. With our freedom to act, we carry out action—that is, our wide range of activities within this world that can only be exercised in a plural environment where other free, acting humans exist. Plurality, then, is the essence of the realm that is shared with others, and it is also the external condition that needs to be met by the "world" or the "space" in which freedom and action can appear. In the next subsection, I elaborate on freedom, which accompanies action.

Freedom

Whenever *praxis* appears, it is always accompanied by the appearance of freedom, which is the capacity to take personal initiative. To grasp *praxis* in its entirety, it is essential to understand its accompanying feature. As already discussed, Arendt characterizes politics as interaction: In politics, human beings meet and take care of their shared world in a public sphere, and politics is based on a certain understanding of commonality while allowing everyone to have their own perspective related to the world that they share. Freedom enables political interaction, allowing individuals to explore diverse perspectives on their shared reality (Canovan, 1994). Below, I explore the various aspects of freedom as it was conceptualized by Arendt.

a) Freedom from Necessity (I-Can) Versus Inner Freedom (I-Will)

In Arendt's terminology, freedom does not come across as inner freedom, free will, or freedom of choice, but it rather means, having the competence to act. It is the capacity to bring something into being, the capacity to begin, and thus, the capacity to carry out *praxis*. Before Christianity, Arendt argues, freedom was located in the human capacity of "I-can," which was a status of the bodily existence of the human instead of a status of the mind (Arendt, 1981a, p. 19). She continues,

> Freedom meant that one could do as one pleased, forced neither by the bidding of a master nor by some physical necessity that demanded laboring for wages in order to sustain the body nor by some somatic handicap such as ill health or the paralysis of one's members. According to Greek etymology, that is, according to Greek self-interpretation, the root of the word for freedom, *eleutheria*, is *eleuthein hopos ero*, to go as I wish, and there is no doubt that the basic freedom was understood as freedom of movement. A person was free who could move as he wished; the I-can, not the I-will was the criterion.
>
> (Arendt, 1981a, p. 19)

This freedom of movement and the capacity of "I-can," refer to a condition in which humans are not driven by necessity. Arendt argues that freedom, free will, and inner freedom, far from being identical, are separate things (Arendt, 1960a).

She traces the first dislocation of the different senses of freedom to the Apostle Paul:

> When we come to Paul the accent shifts entirely from doing to believing, from the outward man living in the world of appearances (himself an appearance among appearances and therefore subject to the semblance and illusion) to an inwardness which by definition never unequivocally manifests itself and can be scrutinized only by a god who also never appears unequivocally.
>
> (Arendt, 1981a, p. 67)

This retreat from the outward world to the inward space of the self was caused by the law of God, which, as Paul understood it, had to come from within. It demanded as a "voluntary submission an I-will of agreement" (Arendt, 1981a, p. 68). This means that doing good is not enough to be good; one has to will good. This faculty, namely willing, meant that one can will or nill to do, and thus, even when the mind tells the body to do, one has the capacity to will or will not (Arendt, 1981a). For Arendt, Paul conceived this willing as an internal struggle of the agent, since sinning, which is the greatest disobedience, potentially exists in everyone. It is the will that stops one from disobeying (Arendt, 1981a).

This discovery of willing later became more significant with St. Augustine's formulation of the concept as freedom of choice and inner freedom. It was with St. Augustine, according to Arendt, that the exercise of free will turned into something that portrays authenticity in humans, rather than the exercise of free will being the exercise of *praxis* (Arendt, 1981a). The work of St. Augustine, Arendt claims, aimed at separating freedom from politics and arrived at the formulation that "one may be a slave in the world but still be free" (Arendt, 2006a, p. 146). From St. Augustine onward, the will became an organ of self-liberation, and what the aim of the will executes became liberating the self from worldly affairs (Hansen, 1993). Here, freedom is explained as the "ability of the self to realize fully his willed aims and intentions" (Dossa, 1989, p. 83). This equation of freedom with free will, which would lead to inner freedom, leads to an equation of freedom with sovereignty, in the sense that free will is something that every human being has, independent of their circumstances in the world (Dossa, 1989).

Inner freedom, as Arendt perceives it, is "the result of an estrangement from the world" (Arendt, 2006a, p. 145). Inner freedom, in that sense, designates an inward space where humans go to escape from external coercion (Arendt, 1960a). The experiences of inner freedom are derivative, Arendt argues, because they are not realized in relation to others (Arendt, 2006a). This inner freedom was discovered in late antiquity, according to Arendt, by those who had no place in the world to be free (Arendt, 2006a). This also shows that having a private space, like a home, was essential to being free in the ancient city-state. In that home, one could be whoever one wanted to be in the privacy of one's household. Epictetus, according to Arendt, transformed this worldly relational character of freedom—being free from necessity and being free to move in the agora of the *polis* among fellow citizens—into a relation one has with oneself. In the inner realm, having power over someone turned into having power over oneself, a form of power that only belongs to the one who is in that inner realm. Via this construction of freedom, the slaves and servants of the ancient Greek city-state could feel freedom. To be free, they did not need to be free citizens of the *polis*, and they did not need to act (Arendt, 1960a). When freedom started being explained in the form of willing, it stopped referring to a political status of people and began referring instead to an inner disposition (Arendt, 1981a). This inner freedom began to represent the absolute freedom within one's self that is necessary for experiencing freedom (Arendt, 1960a). However, Arendt continues,

> it seems safe to say that man would know nothing of inner freedom if he had not first experienced a condition of being free among others as a worldly tangible reality. We first become aware of freedom or its opposite in our intercourse with others, not in intercourse with ourselves. Before it became an attribute of thought or a quality of the will, freedom was understood to be the free man's status which enabled him to move, to get away from home, to go out in the world and meet other people in deed and word. This freedom clearly was preceded by liberation: in order to be free, man must have liberated himself from the necessities of life.
> (Arendt, 1960a, p. 29)

As can be seen here, Arendt does not necessarily deny the existence of freedom that comes from within. What she points out is that inner freedom cannot be considered relevant to political action. Inner freedom is experienced within oneself; freedom, on the other hand, can only be experienced through carrying out action, by acting in the world. Understanding freedom as a form of "inner experience" of the will would radically individualize the experience of freedom. Freedom would exist as a form of compulsion that would position humans against one another in the pursuit of self-interests (Hansen, 1993).

b) Freedom and Natality

As mentioned above, for Arendt, freedom is not only saying I-will, but also I-will therefore I-can. Here, Arendt's appropriation of St. Augustine's concept of natality is very important. Arendt defines the human conditions of natality and freedom to be almost identical. Natality carries within itself the property of freedom: The power to be able to realize freedom makes freedom appear. In this sense, freedom also has a deep bond with action. Arendt illustrates this bond between natality and freedom by quoting St. Augustine: "[*Initium*] *ergo ut esset, creatus est homo, ante quem nullus fuit* (that there be a beginning, man was created before whom there was nobody)"[7] (Arendt, 1998, p. 177). Arendt argues that, according to St. Augustine, *principium* and *initium* were two separate things. *Principium* meant the beginning of the world, and *initium* meant the beginning of humanity. The beginning meant the "beginning of somebody, who is a beginner himself" (Arendt, 1998, p. 177). Arendt argues: "With the creation of man, the principle of beginning came into the world itself, which, of course, is only another way of saying that the principle of freedom began when man was created not before" (Arendt, 1998, p. 177). Freedom is possessed by humans because their "coming into the world" coincided "with the appearance of freedom in the universe" (Arendt, 2006a, p. 165–6). Because the birth of man is a beginning in itself, Arendt argues, "to be human and to be free are one and the same" (Arendt, 2006a, p. 166). The faculty of beginning was introduced into the world when God created humans (Arendt, 2006a). Similarly, when a newborn baby appears in the world, the baby breaks the routine by virtue of being new, and this begins forming new sets of relationships: The woman and the man who are the parents become a mother and a father, and the appearance of the newborn baby is changing the world as it existed before while not being detached from it. The newborn baby is also an end result of the relationships that already existed in the world: The woman and the man shared a history and had a set of relations before the appearance of the baby.

Here, the discussion arrives at the way in which Arendt forged a union between action, freedom, and natality. Natality, the fact that we are born, carries within itself the faculty of freedom, and this faculty can only appear as such—that is, it can only be carried out in the world—when *praxis* is exercised. This means that political freedom refers to a movement of some sort, which is *praxis*. This is why Arendt argues that when humans carry out *praxis*, it resembles a sort of "second birth," so to speak (Arendt, 1998, p. 176). When *praxis* appears, it is never detached from the world in which it appears; it is highly related to the set of relationships that existed before its appearance. However, the appearance of *praxis* is novel and unique, and therefore it is profoundly new, because with every new beginning, *praxis* starts a new set of relationships and reveals the actor that carries out the *praxis* and the changed world in which it appears as it never was before.

Freedom, for Arendt, "is the reason why people live together in political organization at all; without it, political life as such would be meaningless" (Arendt, 1960a, p. 28). Arendt says, "the *raison d'être* of politics is freedom, and its field of experience is action" (Arendt, 1960a, p. 28). This means that the idea behind politics itself is freedom, and this freedom can only be experienced via *praxis*. Arendtian freedom is the freedom to share the world with others, having the freedom to interact with others and experience the world as a shared common space (Arendt, 2006a).

Referring to the ancient Greek experience of the *polis*, Arendt claims that freedom was understood as something that manifests itself in human action, and in that sense, the life of a free human being needed an audience that could see, judge, and remember the activity (Arendt, 2006b). Arendt continues:

> Only in the freedom of our speaking with one another does the world, as that about which we speak, emerge in its objectivity and visibility from all sides. Living in a real world and speaking with one another about it are basically one and the same ... Freedom to depart and begin something new and unheard of or ... the freedom to interact in speech with many others and experience the diversity that the world always in its totality ... most certainly was and is not the end purpose of politics ... something that can be achieved by political means. It is rather the substance and meaning of all things political. In this sense, politics and freedom are identical.
>
> (Arendt & Kohn, 2005b, pp. xxx–xxxi)

It becomes clear from the quotation above that for Arendt, freedom and action are almost identical; they coappear, and they are interrelated to one another. To be free, one has to act, and to act, one has to be free. Freedom accompanies acting, or, as Dossa argues, it is "an accessory of doing and acting" (Dossa, 1989, p. 82). As mentioned before, by situating freedom as an accessory to action, Arendt transforms an inner-worldly phenomenon into a worldly reality and thus turns it into a political manifestation. As a political phenomenon, then, freedom accompanies action to appear in a world that is essentially a space that includes plurality.

Action

As I have been arguing, plurality and freedom are deeply connected to *praxis*. Plurality is the condition in which *praxis* appears while being accompanied by freedom. Natality carries within itself the character of freedom, which resides in it as a potentiality; when *praxis* is realized, freedom becomes a tangible reality that appears in the world, like a second birth, and this second birth is shared with others, in the plurality of perspectives.

By connecting *praxis* to natality, freedom, and plurality, Arendt sets out *praxis* as an ontological condition. Being human means being born into this world. Being born into this world means establishing new relationships, setting things in motion, and beginning something profoundly new that is nevertheless connected to the set of relationships and the world that existed before one's entrance into the world. Because we are human, we are destined for *praxis*: We are destined to appear, begin, and set things in motion. In short, we are destined to act. Not carrying out *praxis* would mean we deny ourselves our humanness.[8]

The Arendtian definition of *praxis* is an activity exercised by an agent that begins something new. By breaking a pattern, *praxis* realizes freedom as a phenomenon to appear, and it takes place among plural actors. Action, for Arendt, always takes place in this manner; if it does not, then that activity cannot be called action. When action is carried out, it discloses two things: the self and the world. *Praxis* is not *for* something. When a pattern is broken, beginning something new, the end result of this activity is not relevant to the value of the activity itself. In that sense, *praxis* has an intrinsic value: It is an end in itself. What matters is its appearance, not its eventual result.

The characteristics of *praxis* are related to its being a set of phenomena.[9] A phenomenon is what shows itself by appearing to the eyes and the senses, and, as Vollrath and Fantel (Vollrath & Fantel, 1977) continue,

> The phenomena "include" those to whom they appear along with the space in which they occur and which determines the relation between phenomena and those who perceive them. Political events are phenomena in a special sense; one might say they are phenomena per se. For they do not appear within a space integral to themselves, like natural phenomena. Nor is the space in which they appear accorded to them by those who perceive them, as is the case with other

> phenomena such as those of art. The space in which political phenomena occur is created by the phenomena themselves. Or to put it more precisely, it is created by the persons whose acts constitute the political events.
>
> (Vollrath & Fantel, 1977, p. 164)

Arendt's understanding and treatment of praxis are in accordance with phenomena and the space of appearance in which the phenomena occur. The space of appearance comes into being when diverse human agents exercise praxis as a total interplay between people, things, and relationships (Vollrath & Fantel, 1977). Below, I explore the characteristics of action—or phenomena—as set out by Vollrath and Fantel.

Political action, which is an instance of *praxis* or a phenomenon, is an inexplicable, incomprehensible appearance that cannot be explained by causal chains of past events. It comes from within the political actor, since every human life carries within itself the potential of carrying out *praxis*: making it appear and begin something new, setting off a chain of new events, and establishing new relationships. *Praxis*, when it is enacted, is irreversible, meaning that since it establishes relationships and sets in motion a chain of events, what has been started when *praxis* begins cannot be stopped (Vollrath & Fantel, 1977). Our humanness is the primary condition for *praxis*. Because we are humans, we carry the possibility to move and act in the world. That is our freedom: our freedom to act. When we act, we always act in a space that we establish with other acting beings, in human plurality. The fact that *praxis* appears within plurality means that others also appear at the same time that I appear. This means that there are simultaneous appearances of each individual who is in the public space. When I appear, I also see others who appear in front of me. This appearance of myself discloses me to the world, and simultaneously, in my self-disclosure, I also disclose the world as it appears to me. When I share how the world appears to me, I also hear and see how the world appears to others. Below, I explore these two dimensions of action.

a) Action as Self-Disclosure

Praxis, in itself, is a disclosive activity. When it is enacted, it discloses the political actor who carries out the *praxis*, and it discloses the world as it appears to the political actor. Bonnie Honig (1993) argues that Arendt's understanding of action is similar to Nietzsche's. They both do not believe that there is a cause for action. According to Nietzsche, as humans, we believe that we are subjects. Quoting Nietzsche, Honig continues:

> [The belief that we are subjects is] … only owing to the seduction of language (and of the fundamental errors of reason that are petrified in it) which conceives and misconceives all effects as conditioned by something that causes effects, by a "subject" … popular morality also separates strength from expressions of strength as if there were a neutral substratum behind the strong man which was free to express strength or not to do so. But there is no such substratum: there is no being behind doing, effecting, becoming: "the doer" is merely a fiction added to the deed—the deed is everything.
>
> (Nietzsche, 1989, p. I. xiii)

According to Honig, Arendt and Nietzsche agree that "there is no being behind doing" (Honig, 1993, p. 78). In other words, the being is created during the act of doing. However, Arendt limits the appearance of action to the political realm, since only "in the political realm, being and appearance coincide" (Villa, 1995, p. 150). *Praxis*, as an end in itself, is not determined by causes, intentions, justifications, motives, or goals of actors; however, no action is unattended by these things (Honig, 1993). This means that *praxis*, rather than being an expression of the political actor, gives birth to the actor that carries out *praxis* at the same time as it is establishing new relations and realities (Honig, 1993).

Arendt argues that in speaking and acting, "men show who they are, reveal actively their unique personal identities and thus make their appearance in the human world" (Arendt, 1998, p. 179). Villa argues that one can easily interpret this sentence as a confirmation of an "expressivist theory of the self" (Villa, 1995, p. 90). However, he continues in agreement with Honig and argues that Arendt rejects "the expressivist conception of self, which assumes a core self, a basic or essential unity of innate capacities that are expressed, actualized, or concretized in the world of appearances" (Villa, 1995, p. 90). Nietzsche and Arendt, Villa argues, oppose the idea that there is a subject or some form of reality behind appearances (Villa, 1995). According to Villa, the "who" that carries out *praxis*, from Arendt's point of view, is a "multiplicity of conflicting drives, needs, and faculties" (Villa, 1995, p. 90) that are "coextensive with" the actions of this "who" (Villa, 1995, p. 90). *Praxis*, in this sense, comes across as individuating but not subject-centered (Honig, 1993, p. 78). What appears to others who see and hear a person while *praxis* is being carried out shows "who" that person is. While *praxis* discloses who a person is, this disclosure is not in that person's control. Arendt says that who a person is

> can be hidden only in complete silence and perfect passivity, but its disclosure can almost never be achieved as a willful purpose, as though one possessed and could

> dispose of this "who" in the same manner he has and can dispose of his qualities. On the contrary, it is more than likely that the "who," which appears so clearly and unmistakably to others, remains hidden from the person himself.
>
> (Arendt, 1998, p. 179)

From this quotation, it is also evident that for Arendt, a person cannot willfully control who they are because this is left to the judgment and opinion of others. A person also cannot fully understand or know who they are; one can only experience one's "whoness" through one's interactions with others. Only through this appearance can the individual be an authentic human being, since it is only through *praxis* that people distinguish themselves. What is revealed, a person's "whoness," is not a tangible quality, and it is not everything that describes that person; it is one dimension of their full identity, and in every act they reveal a different dimension, or maybe recurring dimensions of it, so that who one is "registers only incompletely on one's fellow participants" (Kateb, 2000, p. 138). Thus, every time someone carries out *praxis*, they develop one aspect of who they are. This appearance of *praxis*, the appearance of who one is, rests on initiative, and to refrain from using this initiative would be to deny one's humanness, according to Arendt (1998). So, as can be seen, Arendt is very clear that *praxis* is an ontological condition: It is what makes us, and therefore everything that is related to the notion of *praxis* has ontological connotations in her political theory—connotations of freedom, plurality, and power.

Who we are and what we are take shape by virtue of our interactions with others. In our interactions, we shape one another, as well as the environment around us. In politics, autonomous agents seek to situate themselves as distinct human beings with particular and idiosyncratic speeches and, above all, actions. For Arendt, politics is continuous discourse and action, seeing and hearing one another, and looking at the world from different perspectives. These different perspectives are present even when a person is alone, imagining the lives of others or forming ideas. Politics is an act that arises from human plurality. It is a form of interaction that shapes our lives.

Only through action, Arendt argues, can we be truly human, because only through action can we show who we are and reveal our unique identities (Arendt, 1998). Action as self-disclosure is focused on the agent-revealing capacity of action. The other mode of action, action as world disclosure, is concerned with the capacity to hold humans together. Within this framework, Arendt highlights our capacity to communicate, debate, and discourse. These two modes of action complement one another: the first mode indicates the act of entering the public realm, revealing who we are, and experiencing our humanness; through this experience, we are able to debate, discuss, and potentially come to an agreement with our fellow human beings, which is expressed by the second mode of action.

b) Action as World Disclosure

Arendt, as already mentioned, distances *praxis* from motives, goals, or intentions. Consequently, one might ask, why would we even do anything, let alone expose ourselves in a public place, if our intentions, motives, or goals are not the driving factor? According to Honig, Arendt sometimes implies that one of the reasons for carrying out *praxis* is—ironically—a need to escape the biological, urgent, and irresistible conditions of the body (Honig, 1993). At other times, Honig continues, Arendt focuses more on the common character of our biological condition, which reaches its peak in modernity, where authenticity, spontaneity, and distinctiveness are excluded (Honig, 1993). What our humanness demands from us is something that we cannot escape, and thus the reason to act, as Honig formulates it, is embedded in our "(agonal) passion for distinction and outstanding achievement" (Honig, 1993, p. 80).

The question then becomes, how and when is this passion instigated? Here, Arendt provides the answer as *amor mundi* (love of the world). In politics, it is not our private concerns or our lives that are at stake; it is the world (Arendt, 1960a, p. 35). The continuity and maintenance of this world depend on the continuous appearance of *praxis*, on the fellow citizens who share the same care for the world and its maintenance (Arendt, 1960a). This care for the world, *amor mundi*, as Vasti Roodt brilliantly explains it, is

> a way of reconciling ourselves to the world by fitting ourselves into it that is to say, by making ourselves at home where we are not. In this regard Arendt's argument is diametrically opposed to the notion that we can only be at home in the world by fabricating which generally means: by destroying and remaking the world in accordance with human needs and interests. Her point, in other words, is not that we can be more at home if only we work harder at making the world conform with our requirements, but rather by choosing to fit ourselves into a world that is not in the first place "for us."
>
> (Roodt, 2008, p. 420)

According to Roodt, loving the world is equivalent to choosing the world as one's home (Roodt, 2008). When we first enter the world, it is not initially our home. With care, we change and shape ourselves in relation to the world and thus make it our home. To be able to show this care and fit ourselves into this world—to be at home in the world—we carry out *praxis*. Via carrying out *praxis*, that is, via engaging in public speaking of words and public performance of deeds, we show our care for the world.

For a person to insert themselves in the public realm because of their love of the world, to carry out their *praxis*, there is one more layer that is actually about the subject. For Arendt, carrying out *praxis* requires courage (Arendt, 1960a). It is easy to reside in the comfort and security of the private, but because we care for this world that we share with others, we act. The courage that this requires, Arendt argues, liberates humans from their immediate worries for their own lives and allows them to act "for the freedom of the world" (Arendt, 1960a, p. 35). I believe what gives us this courage—what also gives us an appeal to our love of the world that comes from ourselves—can best be explained with *Befindlichkeit*.[10] Inspired by Heidegger's term, this term can be translated as a disposition toward the world that we love and fit ourselves into. This disposition that each person has toward and within the world opens the possibility of this disposition being appealed to. Therefore, I believe, because of our dispositions, some principles appeal to us and move us to act, and some do not. While the public realm is essential for *praxis* to appear, there is also a subjective dimension in which, because of one's disposition, one finds oneself moved within the world.

To summarize, political activity (*praxis*, phenomena) can only appear in the world of appearances, the public space. It needs the input and feedback of other citizens, first and foremost to show who we are and how we exist in this space of appearances, and second to simultaneously create a common space of appearances for others to show themselves and to help maintain and conserve this space. *Praxis* is principled action: We show what matters to us through our *praxis*, for the love of this world we share with others. Action is a manifestation of our freedom because we can do what we will: We set things in motion, beginning something new or out of the ordinary because we want to, because we care about this world. Arendt says

> Action and speech go on between men, as they are directed toward them, and they retain their agent-revealing capacity even if their content is exclusively "objective," concerned with the matters of the world of things in which men move, which physically lies between them and out of which arise their specific, objective, worldly interests. These interests constitute, in the word's most literal significance, something which *inter-est*, which lies between people and therefore can relate and bind them together. Most action and speech is concerned with

> this in-between, which varies with each group of people, so that most words and deeds are *about* some worldly objective reality in addition to being a disclosure of the acting and speaking agent.
>
> (Arendt, 1998, p. 152)

This world that we share with others, this world that we have love and care for, relates us to one another and separates us from each other (Arendt, 1998, p. 242). It relates us because we are all concerned about the maintenance and security of this world, and it separates us because it allows us to show how it appears to us differently. This world that we care for is a durable, stable place that stores our appearances. Our fellow citizens see and hear us; they register these moments of illumination of principles in their memories and transfer them to further generations, and thus they immortalize us—they immortalize what appeared. These appearances, secured by the durable, tangible world, make us who we are, both as a community and as a nation.

As can be seen, freedom, plurality, and action are very closely related to one another: Through our carrying out *praxis*, we make freedom appear within the plurality of fellow citizens. There is one more important component in Arendtian political theory, which is power. In the next section, I address Arendt's concept of power.

Power

Arendt points out that, within contemporary political science, there is a misguided conception of what *power* really is. According to Arendt, it is important to distinguish between *power*, *strength*, *force*, *authority*, and, finally, *violence* (Arendt, 1970b). All of these different phenomena refer to different forms of relations, but somehow, they tend to be used almost interchangeably in most writing about political science. According to Arendt, "power corresponds to *the human ability* not just to act, but *to act in concert*"[11] (Arendt, 1970b, p. 44). It should be noted that she does not define power in terms of relations of domination. Her definition of power is in contradiction with the mainstream definition of power as domination or *power over*. As mentioned in the introduction, Arendt's concept of power can be defined as *power to* (Haugaard, 2012). According to Göhler, this form of power can be understood as a disposition or a capacity that is latent and invisible when not exercised (Göhler, 2009). Arendt also underlines that power is strongly separated from other concepts that people tend to use interchangeably with it. According to Arendt,

> Strength unequivocally designates something in the singular ... *Force* ... should be reserved, in terminological language, for the "forces of nature" or the "force of circumstances," that is, to indicate the energy released by physical or social movements. *Authority* can be vested in persons—there is such a thing as personal authority, as for instance, in the relation between parent and child, between teacher and pupil—or it can be vested in offices, as, for instance, in the Roman senate or in the hierarchical offices of the Church. (A priest can grant valid absolution even though he is drunk.) Its hallmark is unquestioning recognition by those who are asked to obey; neither coercion nor persuasion is needed. *Violence* is distinguished by its instrumental character. Phenomenologically, it is close to strength, since the implements of violence, like all other tools, are designed and used for the purpose of multiplying natural strength.
>
> (Arendt, 1970b, pp. 44–6)

Furthermore, while strength, force, and authority are measurable and materially bound, power exists only "potentially" (Arendt, 1998, p. 200). Highlighting its etymology, Arendt refers to power as "potential" because the term refers to the "Greek verb of *dynamis* like the Latin *potential* with its various modern derivatives or the German *Macht* (which derives from *mogen* and *moglich,* not from *machen*)" (Arendt, 1998, p. 200).

Potentially, power always exists; however, it only becomes an actuality when people agree to act together. Thus, power needs the appearance of *praxis* to be generated. Only when citizens act out of their care for the world, illuminate principles, and show themselves and how the world appears to them—only then can power be generated. In this sense, power, according to Arendt, is also very materially bound, since the possibility of its emergence is based on individuals appearing to one another in the public space. Power in this sense is conceptualized as productive and is generated through reciprocal interaction. This power, however, can only be maintained as long as people stay together. The moment they disperse and cease acting together, the power that is generated from this togetherness also disappears (Arendt, 2006b). So power actually is the end result of "free action by equals who act in concert, bound together by mutual promises and reciprocity for the sake of bringing something new into being" (Honig, 1993, p. 101).

As can be seen, Arendt does not consider power to be in the possession of an individual but rather to be in the possession of a group—in other words, a community of individuals. Hence, Göhler argues that in Arendt's formulation, power is empowerment: It is "people gaining the ability of autonomous action" (Göhler, 2009, p. 32).

Until now, I have discussed how the notions of plurality, freedom, action, and power are closely related. According to Arendt, it is only when these notions come together that politics occurs, meaning the world is illuminated as politically meaningful. To sum up, plurality is an external condition, an ethos so to speak, that a public space embodies for free agents. These agents have the freedom to act, to be able to enter, to distinguish themselves as distinct, unique beings and thus establish their identity, and also reveal the world as it appears to them to their fellow human beings. When there is an agreement on how the world appears and there is not only reaction to action but also coaction, power is generated to move forward in founding a new beginning, with the ethos of plurality. Thus, politics for Arendt is characterized by *interaction*: in politics, human beings meet and take care of their shared world in a public sphere.

Conclusion

In this chapter, I have demonstrated that the power-oriented characterization of politics evolved from a misconstrued understanding of the basis of human political action. Following Hannah Arendt (Arendt, 1998), I showed that the power-oriented conceptualization of politics actually treats politics as if it were technology. I showed that politics has been coupled with an instrumental, power- and goal-oriented approach toward the world that is inherent in *poiesis*, our human capacity that is guided by *techne*. I have shown that once politics is construed as being interaction-oriented, this opens room for a thorough exploration of the influence of technology in politics. Arendt's political theory, by connecting human plurality, freedom, natality, and power through the concept of *praxis*, revitalizes the public quality of politics. It situates human existence with an ontological connection to the world that can only be meaningful when the world is revealed in the manner of politics. In politics, human beings meet and take care of their shared world in a public sphere, based on a certain understanding of commonality while allowing everyone to have their own perspective related to the world that they share. Politics manifests in between and among human relationships, and therefore, it is interactional rather than technological.

Once the distinction between technology and politics is clearly defined, then the intricate relationship between politics and technology can be investigated without confusing one with the other. While agreeing with Arendt that maintaining the distinction between *poiesis* (that has a result manifested as technology) and *praxis* (that has a result manifested as politics) is essential, I believe Arendt missed an important aspect of *poiesis*: the fact that it has a conditioning and contextualizing the impact on how we exercise our political freedom and express ourselves as the authentic, distinct beings that we are. Even though *poiesis* is an instrumental perspective and not political, without it we would not be able to build the world that we live in. In the next chapter, I show that *praxis* and *poiesis* are intertwined, rather than opposed as Arendt suggests.

Notes

1 Like Arendt, I use *praxis* and action interchangeably, and both terms always refer to political action.

2 Arendt here uses the concepts *bios theoretikos* and *vita contemplativa* interchangeably. The validity of this form of use is open to debate, but it is not in the scope of this book to discuss the matter. For more information on *bios theoretikos* see Bénatouïl & Bonazzi (2012); Cavarero (2002); Leonard (2019); Parekh (1981); Warren (2004).

3 According to Arendt, *bios* (human life) is distinguished from the life of all other entities in the world (Arendt, 1998, p. 97). The appearance and disappearance of *bios*, Arendt says, "constitute worldly events which ultimately can be told as a story, establish a biography" (Arendt, 1998, p. 97). Arendt, as will be seen, throughout her analysis of the activities that constitute *vita activa*, sets action (*praxis*) apart from the other activities. The reason for this is that, in coupling action with the realm of plurality, she finds the perfect definition of Aristotle's *bios politikos*. Arendt's motive in this coupling is to show that freedom cannot be found in contemplation, as was portrayed by the Western philosophical tradition (Villa, 1995).

4 Here, Arendt refers to Homer for the reader to see the use of *prattein* and *archein* (Arendt, 1998, p. 189).

5 Existentialia is a set of existential structures that Heidegger develops in his quest to answer the question "what is being?" In order to understand existentialia, it is first necessary to understand Heidegger's use of the term existence. According to Inwood (1999), existence in its traditional meaning refers to the "existence of an entity, in contrast to its essence" (Inwood, 1999, p. 60). Heidegger's use of the term,

however, applies only to *Dasein* (Inwood, 1999, p. 60). This means that *Dasein* does not have a nature or an essence, as other entities do; "the essence of Da-sein lies in its existence" (Heidegger, 1997, p. 40). According to Heidegger, *Dasein* does not exist in the same way that other entities do, and in that sense the fundamental characteristics of its existence should be termed *Existenzialien*, "existentialia, existentials" (Inwood, 1999, p. 61), rather than "categories."

6 This is related to the inherent characteristics of political action as *praxis*. For Arendt, action (*praxis*) displays three characteristics: it is spontaneous, it is boundless, and it is unpredictable (Arendt, 1998). The spontaneity of action comes from its nature of breaking the routine. Thanks to the fact that it sets something into motion, action breaks the cycle of whatever was happening at the moment of its occurrence, and because it is a beginning, it is new and out of the ordinary. These characteristics of *praxis* make the concept fragile and alarming for one who seeks order and predictability in political life as such.

7 Arendt's translation.

8 Arendt says, "A life without speech and without action, on the other hand—and this is the only way of life that in earnest has renounced all appearance and all vanity in the biblical sense of the word—is literally dead to the world; it has ceased to be a human life" (Arendt, 1998, p. 176).

9 Here I use phenomenon, *praxis*, and action interchangeably since, for Arendt, they are identical.

10 According to Baugh (1989), "Heidegger groups Befindlichkeit … with an understanding and discourse as a form of closedness, which itself is the basic form of truth" (Baugh, 1989, p. 124). The term is often translated as "attunement," "affectivity," or "disposition" (Elpidorou & Freeman, 2015; Loidolt, 2018). In this book, I adopt the translation Loidolt also uses extensively in her work, *Befindlichkeit* as disposition.

11 Arendt's italics.

2

Arendt, Phenomenology, and Technology

In the previous chapter, by examining the main concepts that make up Hannah Arendt's political theory, I argued that Arendt's main concern is to regenerate the original meaning of political action as *praxis*. The displacement of *praxis* (political action) by *poiesis*, according to Arendt, is what leads to the instrumentalization of politics. This displacement entails a change within the hierarchy of the activities of *vita activa*. Observing this change, Arendt argues that humans became alienated from the world that they live in. However, whereas Arendt saw *praxis* and *poiesis* as being opposed to each other, I argue that they are in fact intertwined. In this chapter, I develop this argument in four sections.

In the first section of the chapter, I show that a close inspection of the concepts of *praxis* and *poiesis*, for Arendt, necessitates a close examination of her method of analysis. Phenomenology, specifically hermeneutic phenomenology, informs Arendt's political theory in its entirety: In fact, Arendt's political theory is a hermeneutic phenomenology of the political (Borren, 2013). This observation leads to grasping Arendt's notions of common sense and common world as the building blocks of her political theory. The fact that Arendt positions common sense as both "world building" and "world disclosing" (Borren, 2013) suggests a closer relationship between *poiesis* and *praxis* than Arendt first proposed.

In the second section, I show that Arendt's demarcation of realms within the world indicates that there are physical or material aspects to these realms. I first lay out that artificiality and materiality are intrinsic to the activity of work (fabrication or *poiesis*). Building on this, I show that work, being the activity corresponding to the human condition of worldliness, creates the world. I distinguish Arendt's construction of different realms that the world encompasses. Following these distinctions, I argue that the world Arendt positions as encompassing the realm of the private and public is indeed artificial but also material.

In the third section, I show that Arendt's insistence on opposing *praxis* and *poiesis* stems from her observations regarding modern technology and mass society. Actually, Arendt is very well aware of the influence of the material end result of *poiesis* on *praxis*. However, her reading of this influence is rather pessimistic, since she believes that rapid technological advancements have drastically changed the hierarchy of *vita activa* by placing labor above action. For Arendt, this shift is influential in the emergence of mass society and the creation of a social realm that is neither private nor public but a realm where private matters are exposed in public. Accompanied by the mode of worldlessness, for Arendt, the social realm is the hallmark of mass society. The main problem with Arendt's analysis of modern technology is that she situates technologies within the realm of the social and does not consider their political capacity. This leads her to make observations about modern technological artifice that are similar to those of classical philosophers of technology, such as Karl Jaspers (2014) and Martin Heidegger (1954, 1962, 1997).

In the fourth section, I show that Arendt does not consider all end results of *poiesis* as irrelevant to politics. I argue that if Arendt's conceptualization of the cultural end products of *homo faber*, such as stories and artworks, is considered as a point of departure—rather than her limited reading of technologies—she can guide us to reach an interaction-based politics of technology. Following this, I show that the cultural products of *homo faber* not only stabilize the appearances of *praxis* but also inspire further appearances of *praxis* because they are an embodiment of the meaning that the world has been given at their time of creation. I end the section by arguing that it is postphenomenology, rather than phenomenology, that can help Arendt's political theory to expand the implications of this influence of *poiesis* on *praxis*.

Hermeneutic Phenomenology of the Political

Arendt, in her political theory, argues for a strict demarcation between the realms of action for *poiesis* and *praxis*. To make this demarcation seem as strong as possible, as explained in the previous chapter, Arendt positions *praxis* and *poiesis* as opposed to one another. However, I argue that *praxis* and *poiesis* are, in fact, intertwined rather than opposed. To argue this point, I must first explore the hermeneutic phenomenological method that informs Arendt's political theory.

Hermeneutic Phenomenology

Phenomenology, especially hermeneutic phenomenology, shapes Arendt's political theory in its entirety: In fact, Arendt's political theory is a hermeneutic phenomenology of the political (Borren, 2013). This is because of the phenomenological aspects that can be found in Arendt's method of analysis. The first is her understanding of phenomena and lived experience, which always have political connotations for Arendt. Phenomena are always connected to the artificial world that is the end result of the work of *homo faber*. The second aspect is her dismissal of the subject–object dualism and her formulation of understanding (i.e., making sense of the world) as an end result of lived experience rather than a cognitive endeavor. The third aspect of her hermeneutic phenomenology is her concept of the world itself, which is inspired by Heidegger but adjusted in a way that takes into account human plurality as a component of this world (Borren, 2013). Examining these components of Arendt's hermeneutic phenomenology allows one to grasp the centrality of the artificial world in Arendt's political theory.

The first aspect of Arendt's hermeneutic phenomenology can be found in Arendt's analysis of lived experience (Borren, 2013). According to Borren, Arendt's main interest is in understanding this lived experience in the form of "political phenomena, events and experiences," particularly those that occur in public spaces (Borren, 2013, p. 232). Political phenomena that Arendt is interested in refer to occurrences produced by human actions and interactions (Borren, 2013). For Arendt, phenomena always take the form of the "appearing-of-someone-or-something-to-others" (Borren, 2013, p. 232). According to Borren, Arendt assumes a phenomenological analysis of lived experience in order to understand what this political phenomenon means (Borren, 2013). Here, the meaning of political phenomena is not construed as a "cognitive or epistemological category of intelligibility" by hermeneutic phenomenologists, according to Borren, since intelligibility can only be derived from primary meaningfulness (Borren, 2010, p. 20). From a hermeneutic phenomenological point of view, "meaning refers to meaningfulness," as it is "meaningful contexts or situations in which human life is always embedded" (Borren, 2010, p. 20). Following Arendt's reasoning, Borren explains lived experience as the way in which humans live their lives, carry out projects, and interact with one another (Borren, 2013). Things make sense to us because we deal practically with them, and through these practical dealings with them, we become familiar with them

(Borren, 2013). Experience consists of these practical dealings with things; it assumes a "pre-reflective understanding of phenomena" (Borren, 2013, p. 233). Therefore, we have experiences because things make sense to us, and things make sense to us because we have experiences (Borren, 2013).

The second aspect of Arendt's hermeneutic phenomenology concerns her dismissal of the subject–object dualism. According to Arendt, in order to understand the meaning of an experience, one needs to interpret and disclose what the experience meant, since this meaning is not clear to the one undergoing the experience immediately (Borren, 2013). This means that there is no subject that observes objects; rather, there are phenomena, appearances, objects, or events that are perceived by the perceiver (Borren, 2013). Rejecting the subject–object dualism[1]—in other words, rejecting a consideration of things and events as external to us humans—hermeneutic phenomenology construes phenomena as "always immediately but implicitly meaningful, constituted by implicit understanding based on our familiarity with the world and our know-how" (Borren, 2013, p. 233). Rejecting the subject–object dualism entails that subjects and objects should always be analyzed in their relationship to one another, in the way in which things and events manifest themselves in our lived experiences (Borren, 2013). According to hermeneutic phenomenologists, Borren argues, humans are considered as interpretative beings who find meaning in their encounters and experiences and are inclined toward understanding (Borren, 2013). Interpretation, which is necessary for grasping the meaning of phenomena, depends on understanding (Borren, 2013). In the hermeneutic phenomenological tradition, understanding is an essential mode of being that humans inhabit (Borren, 2013). Understanding in that sense is not something that we do; it is the way in which humans experience things, events, people, and themselves (Borren, 2013).

For hermeneutic phenomenologists, because humans' mode of being is understanding, and because phenomena are always implicitly meaningful, humans experience something-as-something: for instance, a natural event is experienced as a storm, or a political event is experienced as a revolution (Borren, 2013). This experiencing something-as-something is based on humans' "intimate familiarity with and know-how of the world" (Borren, 2013, p. 234), and it is embedded in their "practical dealings with the world and with other people" (Borren, 2013, p. 234). According to Borren, Arendt's method is exactly this analysis of phenomena, such as "political events, through the shared, intersubjective and worldly, experience of these phenomena" (Borren, 2013, p. 234). In that sense, events become the center point of analysis for Arendt, since they illuminate the meaningfulness of the world, and that is exactly what we experience: illumination (Borren, 2013). These illuminations all require interpretation for us to grasp their full meaning, since they are all disclosed parts of reality, and (as mentioned before) we do not immediately know the meaning of these disclosures: They need inquiry and interpretation (Borren, 2013).

The third aspect of Arendt's method that needs attention is her concept of the world. Through that concept, too, Arendt proves to be a hermeneutic phenomenologist because she uses the "notion of world as the meaningful context within which human existence enfolds" (Borren, 2013, p. 235). As I outlined in Chapter 1, Arendt assumes that "human existence is worldly existence," meaning that "we are situated in the world," we are conditioned by the world, and we condition the world in a specific way: in other words, "we are both shaping and shaped by the world, which is public, visible and common to all" (Borren, 2013, p. 235). Here, Borren argues that Arendt is clearly indebted to Heidegger's early work, in which he defines human existence as "being-in-the-world" (Borren, 2013, p. 235). Since Arendt's concept of the world is an interpretation of Heidegger's, it is essential here to turn to Heidegger and grasp how he conceptualized the world.

Heidegger configures the human as *Dasein*, whose "primary existential way of being or of relating to the world, others, and itself (expressed in the so-called 'existentials'), is *in-der-Welt-sein*, being-in-the-world" (Borren, 2013, p. 235). *Dasein*, according to Heidegger, as per Villa's terminology, is a "with world" (*Mitwelt*), which means that there is no human existence that is outside of the world. This means that *Dasein* is always in the world; in other words, *Dasein* is being-in-the-world (Villa, 1995). As Villa puts it,

> in clarifying Being-in-the-world we have shown that a bare subject without a world never "is" proximally, nor is it ever given. And so in the end an isolated "I" without others is just as far from being proximally given. The "others"—the knowability of which had been such a problem from Descartes to Husserl—"already are there with us": we are always already *with* others, just as we are always alongside entities. Considered from the perspective of the "who" of *Dasein*, the world of Being-in-the-world is a "with world" (*Mitwelt*); "Being-in," according to Heidegger, is a "Being-with-others."
>
> (Villa, 1995, p. 122)

What Heidegger aims to highlight with his configuration of human existence as embedded in the world is that humans are part and parcel of the world, which constantly conditions the way in which they exist. This indicates two forms of others that are both influential in human existence since they are both *within the world*: the other as things (including material objects and infrastructures) and the other as other humans. This world that we live in shapes and changes us, because both other humans with whom we share the world and the material objects that are in the world are influential in the way we exist. This world is the ground for meaning for us. For Heidegger, *Dasein* is concerned with disclosing the world by default. As Villa puts it,

> *Dasein*'s originary disclosedness—its having a "world"—is accomplished by a projective understanding that "clears" the world, and orients itself within the horizon of significance, in terms of *Dasein*'s existential possibilities or "potentiality for Being." As disclosedness, *Dasein* is a projective structure of understanding, an open structure of possibility ... *Dasein* is constantly at work bringing new "entities"—vocabularies, practices, beliefs—to light.
>
> (Villa, 1995, pp. 125–7)

According to Heidegger, the world is an ensemble not of things but of relationships, an extensive network of instrumental or equipmental relationships (Villa, 1995). For Heidegger, Dasein is thrown to the world, which emphasizes *Dasein*'s entanglement in the world (Topolski, 2015). The fact that *Dasein* is in the world is not a choice for *Dasein* since *Dasein* is always already in the world; it is thrown or projected (*geworfen*) into the world and defined by its possibilities and limitations, such as death (Topolski, 2015).

Topolski argues that, for Arendt, the most important aspects of the Heideggerian notion of the world are worldhood (*weltheit*):

> which is an existential and ontological notion, that can be best understood as the background or horizon upon which all things appear, and world disclosure (*erschlossenheit*), which is how things (including the other for Heidegger) become meaningful or appear to *Dasein*.
>
> (Topolski, 2015, p. 17)

According to Topolski, Arendt diverged from Heidegger when she recognized that even with his focus on history and being-in-the-world, he failed to "recognize that the other cannot be understood or reduced to an object or product of *Dasein*'s existence" (Topolski, 2015, p. 14). Arendt thus moved away from prioritizing the self and moved toward "the other" and intersubjectivity (Topolski, 2015, p. 17). This idea of being-with-others informs Arendt's construction of the world, which is shared with others in the manner of

Miteinandersein (being-with-others), as the ground in which the self discloses itself and appears to others (Topolski, 2015, p. 19).

According to Borren, by accepting Heidegger's understanding of being human as being-in-the-world but emphasizing being-in-the-world-with-different-others, Arendt highlights the *plural* worldly aspect of human existence (Borren, 2013). For Borren, Arendt's hermeneutic phenomenological method has implications for her view of the common world and common sense (Borren, 2013). By highlighting the plural worldly aspect of human existence, Arendt construes reality as a common context that is shared with others. In other words, Arendt argues that humans cannot perceive reality in isolation from others. Reality always becomes illuminated, perceived, understood, and judged within our relation to others and our shared common world. Our common world, however, can only be translated into a communicable context through common sense.

Common Sense

According to Arendt, common sense is "our mental organ for perceiving, understanding, and dealing with reality and factuality" (Arendt, 1970b, p. 8). For Borren, this means that what we humans feel as real has nothing to do with the fact that we have a common nature, but everything to do with the fact that we have common sense (Borren, 2013). With this distinction in mind, Remi Peeters (2009a) argues that common sense in Arendt's terminology refers to "two related, yet different faculties" (Peeters, 2009a, p. 337). Below, following Peeters' analysis, I describe these two different formulations of the common sense that informs Arendt's political hermeneutics.

a) Common Sense as Sixth Sense

The first faculty that common sense refers to in Arendt's use of the term is *the sixth sense*, which designates a cognitive faculty. However, this cognitive faculty should not be confused with thinking; the former is embedded within the world, whereas thinking necessitates a distance from the world. According to Peeters, Arendt's analysis of common sense as the sixth sense can be summarized in two propositions: "first, there is no truth without common sense; second, thinking as such is never able to establish any form of knowledge or truth" (Peeters, 2009a, p. 347). Below, following Peeters, I unpack these propositions.

The first proposition, that there is no truth without common sense, is related to Arendt's perception of the world as the basis for lived experiences. As mentioned before, the world, for Arendt, is a world of appearances. In this world of appearances, perception, even though it is subjective, is essential, but this subjective perception cannot be completely relied upon. In *The Life of the Mind*, Arendt writes:

> Seeming—the it-seems-to-me, *dokei moi*—is the mode, perhaps the only possible one, in which an appearing world is acknowledged and perceived. To appear always means to seem to others, and this seeming varies according to the standpoint and the perspective of the spectators.
>
> (Arendt, 1981a, p. 21)

However, referring to Merleau-Ponty, Arendt argues that there is a level of certainty in the perception of the manifestation of things:

> That appearance always demands spectators and thus implies an at least potential recognition and acknowledgment has far-reaching consequences for what we, appearing beings in a world of appearances, understand by reality, our own as well as that of the world. In both cases, our "perceptual faith" as Merleau-Ponty has called it, our certainty that what we perceive has an existence independent of the act of perceiving depends entirely on the object's also appearing as such to others and being acknowledged by them. Without this tacit acknowledgment by others we would not even be able to put faith in the way we appear to ourselves.
>
> (Arendt, 1981a, p. 46)

For Arendt, the realness of what is perceived depends, on the one hand, on the common worldly context and, on the other hand, on the working together of the five human senses (Arendt, 1981a). Arendt describes this interplay between the five senses and the common worldly context:

> This same sense, a mysterious "sixth sense" because it cannot be localized as a bodily organ, fits the sensations of my strictly private five senses—so private that sensations in their mere sensational quality and intensity are incommunicable—into a common world shared by others. The subjectivity of the it-seems-to-me is remedied by the fact that the same object also appears to others though its mode of appearance may be different.
>
> (Arendt, 1981a, p. 50)

Peeters argues that, for Arendt, without the "sixth sense, it would be difficult to explain the sensation of realness" (Peeters, 2009a, p. 342).

The sixth sense guarantees a threefold commonness for Arendt:

> the five senses, utterly different from each other, have the same object in common; members of the same species have the context in common that endows every single object with its particular meaning; and all other sense-endowed beings, though perceiving this object from utterly different perspectives, agree on its identity. Out of this threefold commonness arises the sensation of reality.
>
> (Arendt, 1981a, p. 50)

The five senses, says Arendt, refer to different perceptible sensory properties of the world. Arendt continues: "Our world is visible because we have vision, audible because we have hearing, touchable and full of odors and tastes because we have touch, smell, and taste" (Arendt, 1981a, p. 50). However, in addition to these perceptible sensory properties of the world, there is the sense of realness that corresponds to the aforementioned sixth sense (Arendt, 1981a, p. 50). Here, Arendt says this sixth sense does not sense reality in the strict meaning of the term; rather, the "sensation of reality, of sheer thereness, relates to the context in which single objects appear as well as to the context in which we ourselves as appearances exist among other appearing creatures" (Arendt, 1981a, p. 51). Arendt refers to Thomas Aquinas when she constitutes the sixth sense as the common context:

> a feeling of realness (or irreality) actually accompanies all the sensations of my senses, which without it would not make "sense." This is why Thomas Aquinas defines common sense, his *sensus communis* as an "inner sense"—*sensus interior*—that functioned as "the common root and principle of the exterior senses."
>
> (Arendt, 1981a, p. 51)

According to Borren (2013), common sense integrates our five senses into a context that we share with all human beings; it "fits" us into a common world (Borren, 2013, p. 239). Borren argues that Arendt, by positioning common sense as the sixth sense, shows that human perception of reality and the susceptibility of humans to appearances play "a crucial role in the constitution and the disclosure of the common world" (Borren, 2013, p. 238). Arendt notes that it is tempting to equate this invisible "inner sense" with thinking (Arendt, 1981a, p. 51). However, she is adamant that thinking and the sixth sense are two different things. For Arendt, thinking does not deal with the realm of appearances:

> thinking itself is not only invisible but also deals with invisibles, with things not present to the senses though they may be, and mostly are, also sense objects, remembered and collected in the storehouse of memory and thus remembered for later reflection.
>
> (Arendt, 1981a, p. 51)

According to Arendt, thought processes can be located only in the brain without relying on any biological sensory data, but the feeling of reality belongs to the biological sensory data and common-sense reasoning (Arendt, 1981a). Where common sense has a "matter-of-fact" relation to reality, thinking lacks this relation to reality because it doubts everything it perceives (Arendt, 1981a, p. 52). Arendt continues:

> thinking can neither prove nor destroy the feeling of realness arising out of the sixth sense, which the French, perhaps for this reason, also call *le bon sens*, the good sense; when thinking withdraws from the world of appearances, it withdraws from the sensorily given and hence also from the feeling of realness, given by common sense.
>
> (Arendt, 1981a, p. 52)

Peeters argues that although it is true Arendt positions the sixth sense as opposed to thinking, knowledge, and science are not the opposite of this version of common sense (Peeters, 2009a). According to Peeters, Arendt makes a distinction between thinking and knowledge: Whereas "thinking asks questions of *meaning*, the goal of knowledge is irrefutable *truth* with respect to reality" (Peeters, 2009a, pp. 344–5). This does not mean that Arendt disregards the role of thinking in arriving at truth, but she believes that thinking alone is not adequate for arriving at that irrefutable truth (Peeters, 2009a). Peeters claims that according to Arendt, "truth must appear, reveal itself, and it does so in the first instance to the senses" (Peeters, 2009a, p. 345). Arendt goes on: "cognition, whose highest criterion is truth, derives that criterion from the world of appearances in which we take our bearings through sense perceptions, whose testimony is self-evident, that is, unshakeable by argument and replaceable only by other evidence" (Arendt, 1981a, p. 57). Here, Peeters argues, Arendt shows that "the roots of truth as a criterion for knowledge" (Peeters, 2009a, p. 345) are found first in sensory perception, which makes knowledge and operates in the domain of common sense (Peeters, 2009a). According to Peeters, common sense, which Arendt speaks of as the sixth sense, is a cognitive faculty that is "a precondition for every experience and for all knowledge of reality" (Peeters,

2009a, pp. 346–7). Peeters argues that it is this faculty that allows humans to orient themselves in the world of phenomena that they share with others, to observe truth, "to make true statements, and to distinguish true statements from false statements" (Peeters, 2009a, p. 347).

The second proposition, that "thinking as such is never able to establish any form of knowledge or truth" (Peeters, 2009a, p. 347), is closely related to the concept of truth that Arendt finds to be relevant to politics. There is, however, one distinction that needs to be made here. When Arendt opposes truth to the feeling of reality, she is talking about factual truth rather than rational truth (Peeters, 2009a). Rational truth is mathematical or logical truth for Arendt: "even God cannot cause two times two not to make four" (Arendt, 2006a, p. 70). According to Peeters, this form of truth does not belong to the realm of appearances; it is rooted not in "something outside us in the world, which others observe and which thus becomes 'common,' but rather in the laws that govern the functioning of our brains" (Peeters, 2009a, p. 349). Rational truth, in that sense, "enlightens human understanding" (Arendt, 2006a, p. 238). This truth is the form of truth the human being should fall back on under the rule of totalitarianism, when humans experience loneliness due to the nonexistence of public spaces. In *The Origins of Totalitarianism* (Arendt, 2004), Arendt writes:

> The elementary rules of cogent evidence, the truism that two and two equals four, cannot be perverted even under the conditions of absolute loneliness. It is the only reliable "truth" human beings can fall back upon once they have lost the mutual guarantee, the common sense, men need in order to experience and live and know their way in a common world.
>
> (Arendt, 2004, pp. 614–15)

Factual truth, by contrast, belongs to the realm of appearances and thus must inform opinions, according to Arendt. Factual truth, she continues,

> is always related to other people, it concerns events and circumstances in which many are involved; it is established by witnesses and depends upon testimony; it exists only to the extent that it is spoken about even if it occurs in the domain of privacy.
>
> (Arendt, 2006a, pp. 233–4)

Arendt indicates that because "factual truth is always related to other people" and phenomena, "it is political by nature" (Arendt, 2006a, p. 234). Peeters argues that belonging to the phenomenal world brings with it vulnerability. The vulnerability of factual truth is also due to its contingent character: According to Peeters, "it could always have been otherwise" (Peeters, 2009a, p. 349). In that sense, even though factual truth informs opinions, "it is not more self-evident than opinion" (Arendt, 2006a, p. 239). It is essential here to understand that, even though opinions can differ from one person to another, they are only legitimate as long as they are in line with the factual truth (Peeters, 2009a, p. 352).

Through this discussion, I have shown that one of the formulations of common sense according to Arendt is the sixth sense, which she understands as a cognitive capacity (Peeters, 2009a). However, Arendt also formulates common sense as a human capacity to judge (Peeters, 2009a). In the following subsection, I discuss this second version of Arendt's common sense.

b) Common Sense as Sensus Communis

The second description of common sense—as *sensus communis*—is based on Arendt's understanding that making judgments, and the human capacity to judge, constitute a political faculty. This is an understanding that she bases on Kant's *Critique of Judgment*. According to Peeters (2009b), Arendt, in her *Lectures on Kant's Political Philosophy* (Arendt, 1982), portrays common sense "to imply a specific faculty of judgment, which she ultimately designates with the expression 'community sense'" (Peeters, 2009b, p. 411). As mentioned before, Arendt considers human "being" not as being-in-the-world, as Heidegger does, but as being-in-the-world-with-others. Judgment, for Arendt, is an essential human faculty that needs to be exercised by humans, since "judging is one, if not the most, important activity in which this sharing-the-world-with-others comes to pass" (Arendt, 2006a, p. 218). The reason Arendt pays such special attention to judgment is that, according to her, judgments are the basis for human opinions. Humans act politically when they speak their minds, making their opinions about public matters heard by other people with whom they share the world. Through this speaking of opinions, humans "assume their responsibility for the common world they share with others" (Peeters, 2009b, p. 411). According to Peeters, this second definition of common sense departs from Arendt's first formulation of common sense as a cognitive capacity (Peeters, 2009b). The basis of this second description of common sense is in Kant's *Critique of Judgment*. Arendt says,

> In the *Critique of Judgment* Kant insisted upon a different way of thinking, for which it would not be enough to be in agreement with one's own self, but which consisted of being able to "think in the place of everybody else" and which he therefore called an "enlarged mentality."
>
> (Arendt, 2006a, p. 217)

According to Peeters, Arendt considers this enlarged way of thinking as proper to the formation of opinions, since it requires imagining the opinions of people who are absent in order to represent them (Peeters, 2009b). This imagination requires not concerning oneself with one's own interests (Peeters, 2009b). Borren argues that, through imagination, one allows the immediate sensory perception of phenomena to wither away and distances oneself from subjective emotional engagement, thereby enabling critical thinking (Borren, 2013). Through imagination, one reaches this enlarged mentality wherein one imagines other possible perspectives in relation to the phenomena (Borren, 2013). Imagination and enlarged thought enable judgment to be an inherently plural activity, since

> The power of judgment rests on a potential agreement with others, and the thinking process which is active in judging something is not, like the thought process of pure reasoning, a dialogue between me and myself, but finds itself always and primarily, even if I am quite alone in making up my mind, in an anticipated communication with others with whom I know I must finally come to some agreement. From this potential agreement judgment derives its specific validity.
>
> (Arendt, 2006a, p. 217)

For Arendt, this specific validity of judgment cannot extend further than the others that the judging person imagines in his or her considerations (Arendt, 2006a). In that sense, this judgment is community-based, not universal. For Arendt, judgment is political in the sense that it is

> the ability to see things not only from one's own point of view but in the perspective of all those who happen to be present; even that judgment may be one of the fundamental abilities of man as a political being insofar as it enables him to orient himself in the public realm, in the common world.
>
> (Arendt, 2006a, p. 218)

The judgment Arendt takes from Kant here, however, was not developed by Kant for the realm of politics. In his *Critique of Judgment*, Kant was positioning judgment as an aesthetic endeavor. Thus, aesthetic judgment, for Kant, was closely related to the concept of taste (Arendt, 2006a). Arendt was convinced that Kant left the political implications of his concept of aesthetic judgment underdeveloped. She indicates:

> the reason why I believe so much in Kant's *Critique of Judgment* is not because I am interested in aesthetics but because I believe that the way in which we say

> "that is right, that is wrong" is not very different from the way in which we say "this is beautiful, this is ugly."
>
> (Young-Bruehl, 2004, p. 452)

Arendt, then, saw a clear political value in the human judgment of taste. For her, it is also taste that informs people's political judgments, because both taste judgments and political judgments are concerned with the world. According to Arendt,

> in aesthetic no less than in political judgments, a decision is made, and although this decision is always determined by a certain subjectivity, by the simple fact that each person occupies a place of his own from which he looks upon and judges the world, it also derives from the fact that the world itself is an objective datum, something common to all its inhabitants. The activity of taste decides, how this world, independent of its utility and our vital interests in it, is to look and sound, what men will see and what they will hear in it. Taste judges the world in its appearance and its worldliness; its interest in the world is purely "disinterested" and that means that neither the life interests of the individual nor the moral interests of the self are involved here. For judgments of taste, the world is the primary thing, not man, neither man's life nor his self.
>
> (Arendt, 2006a, p. 219)

According to Peeters, it is exactly taste's concern with the world as a space of appearance that makes it political (Peeters, 2009b). However, although it is equally concerned with the world, taste judgment is as subjective as political judgment, according to Arendt, who argues that the "it-pleases-or-displeases-me" (Arendt, 1982, p. 66) is very much present in taste. Taste is immediate, and it is not mediated by any thought or reflection; therefore, the sense of taste is subjective (Arendt, 1982). For Arendt, the biggest problem with taste is that it is noncommunicable (Arendt, 1982). Taste and smell, Arendt says, are entirely private senses, and they are not related to objective things; in other words, they are not necessarily related to objects (Arendt, 1982). The example Arendt gives here is the food one eats and the rose one smells: the sensations that a person attaches to the taste of food and the scent of a rose are inside that person—they are both *inner*[2] senses (Arendt, 1982). One is directly affected by both taste and smell. For that reason, Arendt argues, one cannot debate whether a taste or a smell is right or wrong.

However, there is a solution to this noncommunicability of taste: imagination and common sense (Arendt, 1982). As mentioned before, imagination makes present what is absent (Arendt, 1982). Through imagination, one does not necessarily have to be directly confronted with an object to grasp its sense, and thus "imagination transforms the objects of objective senses into sensed objects, as though they were objects of inner sense" (Arendt, 1982, p. 65). This, for Arendt, happens when one reflects upon a representation of an object rather than the object itself; what causes one's pleasure or displeasure in this scenario is not the direct perception of the object (Arendt, 1982). Arendt indicates that Kant calls this process "the operation of reflection" (Arendt, 1982, p. 65). This operation of reflection, thanks to imagination, creates a distance between the human and the immediate or direct senses that are awakened, which enables one to "mentally" taste what one sees and hears (Peeters, 2009b, p. 418). This distance allows one to be able to make a distinction between what pleases or displeases one in representation when those things are not present for the sensory perceptions (Peeters, 2009b). According to Arendt, this distance means that one can talk about judgments rather than taste, because

> though it still affects one like a matter of taste, one now has, by means of representation, established proper distance, the remoteness or un-involvedness or disinterestedness, that is requisite for approbation and disapprobation, for evaluating something at its proper worth. By removing the object, one has established the conditions for impartiality.
>
> (Arendt, 1982, p. 67)

This distance or disinterestedness is essential for one to be able to make a judgment, since only through this distance can one establish impartiality. It can be concluded, then, that judgment is not possible without imagination, because only the impartiality of imagination guarantees that one can make a judgment.

In addition to requiring imagination, judgment, for Arendt, also requires common sense, or *sensus communis*. According to Arendt, Kant was very well aware that even in the most private and subjective senses, there was something nonsubjective (Arendt, 1982). Referring to Kant, Arendt shows that the human interest in beauty only exists because humans live in a society (Arendt, 1982). The judging of something as beautiful or not does not make sense for Arendt if it cannot be communicated with others; we seek an agreement with the people we share the world with through our judgments of taste (Arendt, 1982). According to Arendt, Kant shows that the fact we share the world with others means that we need to "overcome our special subjective conditions for the sake of others" (Arendt, 1982, p. 67). Arendt interprets the nonsubjective element in the nonobjective senses Kant talks about as intersubjectivity (Arendt, 1982). Arendt explains her interpretation thus:

> Judgment, especially judgments of taste, always reflects upon others, and their taste, takes their possible judgments into account. This is necessary because I am human and I cannot live outside the company of men. I judge as a member of this community.
>
> (Arendt, 1982, p. 67)

According to Arendt, judgment is the end result of two mental operations: the first is the "operation of the imagination, in which one judges objects that are no longer present, that are removed from immediate sense perception" (Arendt, 1982, p. 68), which prepares one for the second operation, that of reflection (Arendt, 1982). Arendt asserts that this "operation of reflection is the actual activity of judging something" (Arendt, 1982, p. 68). The standard for the operation of reflection is its communicability since it helps one to distinguish between approval or disapproval of something that pleases one. The standard on which one bases these decisions is "common sense" (Arendt, 1982, p. 69). Arendt argues that Kant uses the Latin term *sensus communis* instead of common sense because he intends the term to mean "an extra sense—like an extra mental capability—that fits us into a community" (Arendt, 1982, p. 70). *Sensus communis*, for Arendt, is a specifically human sense because human speech and communication depend on it. Arendt quotes Kant to explain further:

> Under the *sensus communis* we must include the idea of a sense *common to all*, i.e., of a faculty of judgment which, in its reflection, takes account (a priori) of the mode of representation of all other men in thought, in order, as it were, to compare its judgment with the collective reason of humanity … This is done by comparing our judgment with the possible rather than the actual judgments of others, and by putting ourselves in the place of any other man, by abstracting from the limitations which contingently attach to our own judgment.
>
> (Kant, *Critique of Judgment*, Chapter 40 as quoted in Arendt, 1982, p. 71)

With this quotation from Kant, Arendt shows that judgment is essentially based on common sense: We are able to imagine because of *sensus communis*. Peeters argues that the fact that judgments are based on common sense does not give them objective validity (Peeters, 2009b). As Arendt says, the validity of these judgments is not the same as the validity of cognitive or scientific propositions, which are based on the evidence of either one's senses or one's mind (Arendt, 1982). However, judgments have subjective validity (Peeters, 2009b) because they are based on the common sense of the community, and they are the end result of reflection that takes one's own judgments and the possible judgments of all others into account (Arendt, 1982). The fact that judgments have subjective validity means that one can never force someone "to agree with one's judgments" (Arendt, 1982, p. 72). According to Arendt, when one judges, "one" can only "woo" or "court" the agreement of everyone else. And in this persuasive activity, one actually appeals to the "community sense." In other words, "when one judges, one judges as a member of a community" (Arendt, 1982, p. 72). According to Peeters, Arendt clearly reads Kant's *sensus communis* as the essential political character of human existence (Peeters, 2009b). Peeters claims that for Arendt, without *sensus communis*, it would be impossible for humans to make themselves understood by others or to share any meaning with one another (Peeters, 2009b). Thus, *sensus communis* enables a political community, because it allows people to speak not only in terms of their needs but also in terms of their judgments (Peeters, 2009b). There would be no common world or a political community if there were no *sensus communis* (Peeters, 2009b). One thing Peeters highlights here is that, while human beings judge as part of a community, they do not stop being worldly. As Arendt says,

> One judges always as a member of a community, guided by one's community sense, one's *sensus communis*. But in the last analysis, one is a member of a world community by the sheer fact of being human; this is one's "cosmopolitan existence." When one judges and when one acts in political matters, one is supposed to take one's bearings from the idea, not the actuality, of being a world citizen and, therefore, also a *Weltbetrachter*, a world spectator.
>
> (Arendt, 1982, pp. 75–6)

In other words, people judge as part of a community and also as part of the world (Peeters, 2009b).

The Relationship Between Sixth Sense and Sensus Communis

It becomes clear that these "two related yet different" definitions of common sense—one as sixth sense and one as *sensus communis*—are essential for the "realization and preservation of a common world" (Peeters, 2009b, p. 431). In sum, sixth sense and *sensus communis* are two perspectives of common sense for Arendt. The sixth sense coordinates one's senses and helps one grasp the realness of one's perceptions. The community sense, as a shared context of meaning, fits people into a community and allows them to communicate their opinions (Peeters, 2009b). By analyzing common sense and the world as they feature in

Arendt's hermeneutic phenomenology, Borren reveals that the world and the notion of common sense are co-original. Common sense can also be understood as the preliminary understanding that humans refer to, which is common to us all, is uncritical, and is inarticulate (Borren, 2013). In that sense, Borren argues that Arendt also situates common sense as the basis for understanding (Borren, 2013). In other words, we not only judge but also understand by referring to common sense.

According to Borren, understanding is rooted in prior implicit understanding (Borren, 2013, p. 40), for instance, in the form of presuppositions. This implicit understanding is something that we reach through our practical dealings with the world; in other words, we are familiar with the world in a basic, intuitive way. Therefore, "we always have an immediate, intuitive, implicit, and nonreflective understanding of the things, events, and other people in the world" (Borren, 2013, p. 240). This preliminary, implicit understanding is sense, and explicit understanding is the articulated and confirmed version of this sense-based preliminary understanding (Borren, 2013). For Arendt, because humans are situated and conditioned beings,[3] the way in which humans relate to one another and to the world always happens in the form of a circle (Borren, 2013). For Arendt, as for other hermeneutic phenomenologists, because understanding and judging are ways in which humans relate to one another and to the world, these two activities also occur in a circular structure (Borren, 2013). Below, I go into detail about what this circular structure entails.

c) The Hermeneutic Circle

This circular way of making sense of the world is called the hermeneutic circle (Borren, 2013). According to Borren, the hermeneutic circle starts when people become engaged by a phenomenon; in other words, it is "only when for other reasons our attention is awakened, that the automatic pilot of everyday, implicit understanding makes a place for the circle of explicit understanding or interpretation" (Borren, 2013, p. 240). For Borren, Arendt articulates taste as the way in which something awakens our attention, since it is taste that enables people to discriminate what appeals to them and what does not in an immediate fashion (Borren, 2013). Taste here works as a "catalyst for the process of acquiring explicit understanding" (Borren, 2013, p. 240). Because a phenomenon appeals to our taste and attracts our attention, it means that the phenomenon affects us, and thus a hermeneutic circle opens (Borren, 2013). Borren argues that "phenomenological analysis of the lived experience of the world" (Borren, 2013,

p. 240) is the starting point of the hermeneutic circle. According to Borren, "this lived experience of the world, … is full of uncritical, pre-reflective, and prejudiced understanding of meaning" (Borren, 2013, p. 240); therefore, we need to "distance ourselves from these prejudices" (Borren, 2013, p. 240) and apply our individual critical reflection.

Borren argues that it is the human capacity for imagination and storytelling that allows someone to create a distance and apply their critical reflection to the phenomenon that attracted their attention (Borren, 2013). By enabling representative thinking, storytelling and imagination allow one to form a more critically reflected version of the implicit understanding that one started with through the attraction toward a phenomenon (Borren, 2013). This critically reflected version of the implicit understanding is explicit understanding or interpretation (Borren, 2013).

According to Borren, explicit understanding or interpretation is a contingent, open-ended, perpetual process: in other words, a hermeneutic circle in which there is constant engagement between the subject and the object, between the "thinking ego" and "the events and phenomena of the world" (Borren, 2013, p. 241). For Borren, this circular relationship between humans and the world implies a codependent relationship between human existence and the world, in the sense that the way in which we understand ourselves and the way in which we understand the world are co-original: they are "mutually conditional" and mutually inform each other (Borren, 2013, p. 241). To reiterate her point, Borren refers to Ramberg and Gjesdal: "[We] simply cannot understand ourselves without the detour through the world, and the world cannot be understood without reference to [our] way of life" (Ramberg & Gjesdal, 2005). Arendt positions common sense as "the source and substance of the hermeneutic circle" (Borren, 2013, p. 241). Although, as mentioned before, common sense includes preliminary and implicit understandings, such as prejudices, it is still the only step toward reflective understanding (Borren, 2013). Even though preliminary understanding includes prejudices and subjective or partial preferences, common sense and imagination together ensure that reflective understanding, in other words, judgment, is not a reflection of preliminary understanding (Borren, 2013).

According to Borren, common sense, by achieving its validity through intersubjectivity, avoids both subjectivism and universalism (Borren, 2013). In sum, common sense, on the one hand, is embedded in the human condition and thus dependent on historical and political context, and on the other hand, it constitutes the context for humans to act, speak, judge, and make sense of the world (Borren, 2013). In other words, common sense is both world building and world disclosing (Borren, 2013). Borren argues that "common sense is co-original with the common world," meaning that "common sense both presupposes and enables a common world and fits us into it," and thus common sense and common world are interdependent (Borren, 2013, p. 248). According to Borren,

> Common sense refers both to a sense we have in common as worldly creatures, that is, as beings who create an artificial habitat to survive, and to a sense of the common, which cannot simply be reduced to the first. It is something which emerges in the space between a plurality of actors and spectators, in our perpetual interaction with the common world, and which maintains this common world at the same time. It is both a feature of the human condition, and, as such, may or may not be realized, and something acquired through socialization in a particular community.
>
> (Borren, 2013, p. 248)

As Borren describes it, then, common sense is Arendt's further reassurance that the world we inhabit is construed by humans as a plural world consisting of different perspectives and different human beings. This being so, common sense lends further emphasis to Arendt's argument against the displacement of *praxis* by *poiesis*. However, if, as Borren argues, common sense is both world building and world disclosing (Borren, 2013), this suggests that common sense is the end result of the interaction between two activities that Arendt constantly positioned as opposites. If common sense is both world building and world disclosing, then *poiesis* and *praxis* must be intertwined rather than opposed. In the next section, I examine *praxis* and *poiesis* closely in relation to their common denominator, the world.

Common World and Worldliness

In this section, I first show that artificiality and materiality are intrinsic to the activity of work (fabrication or *poiesis*). Building on this, I show that work, being the activity corresponding to the human condition of worldliness, creates the world. I distinguish Arendt's construction of different realms that the world encompasses. Following these distinctions, I argue that the world Arendt positions as encompassing the realms of the private and public is indeed artificial but also material.

One thing that Arendt constantly reminds her readers is that human life is composed of modes of conditioned existence. As I described in the first chapter, there are different modes of existence within life, and one of the conditions that shapes the artificial existence of humans is the condition of worldliness. The activity that corresponds to worldliness is work, which is *poiesis*. According to Arendt, human life as it is would not be possible if things, which are the end result of *poiesis*, were not involved (Arendt, 1998). These human-constructed things come together and make up the world that we live in. Loidolt (2018) provides her own translation of parts of the section on "work" from the German version of *The Human Condition*, since some parts related to the concept of work were not translated into the English version. These parts are important for the purposes of this book to show how and why the Arendtian notion of work is essential for her political theory and also for the political philosophy of technology. Loidolt's section on work reads as follows:

> Work provides an "artificial" world of things, distinctly different from all natural surroundings [the German version adds: *in that they withstand nature to a certain degree and are not simply powered by life processes*[4]]. Within its borders each individual life is housed, while this world itself is meant to outlast and transcend them all [the German version adds: *in this world of things human life is at home, which is by nature homeless; and the world offers a home to humans to the extent that it outlasts human life, withstands it and confronts it objectively*] the human condition of work is worldliness [the German version adds: *namely the dependence of human existence on objecthood* (Gegenstandlichkeit) *and objectivity*].
>
> (Loidolt, 2018, p. 109–10)

In the English version of *The Human Condition*, the definition of the condition of worldliness, namely the dependence of human existence on objecthood, is excluded. As can be seen, for Arendt, there is a close relationship between humans and things. She argues:

> The objectivity of the world—its object- or thing-character—and the human condition supplement each other; because human existence is conditioned existence, it would be impossible without things, and things would be a heap of unrelated articles, a non-world, if they were not the conditioners of human existence.
>
> (Arendt, 1998, p. 9)

It is clear, then, that Arendt celebrates these artificial things we create to house us to provide shelter for us, and she also recognizes that there is a close relationship between these things we create and how we exist: They condition our very existence. The public world that we occupy as humans, according to Arendt, is not like nature or earth; it is a product of human artifact, the result of the "fabrication of human hands" (Arendt, 1998, p. 52). According to her, living in the world means that the world itself is in between separate human individuals or communities; there is always something that is common to humans in material form (Arendt, 1998, p. 52). Pieter Tijmes (1992) argues that Arendt is indeed an advocate of the artificial character of humanity. According to Tijmes, for Arendt, this capacity to create the artificial, to fabricate, is what is specific about being human (Tijmes, 1992).

The fabrication Arendt celebrates is the human condition of "*worldliness.*" Canovan (Canovan, 1994) observes that work, even though it is not considered a public activity by Arendt, has a strong affinity with the public realm. This affinity stems from the fact that work produces material things, and these things exist in the world, which is shared by everyone (Canovan, 1994). Villa argues that it is work that gives humans a world consisting of a realm of objects, and this world is not a heap of unrelated artifacts or things, but rather they are categorized in terms of means and ends (Villa, 1995). This world that *homo faber* builds, the human artifice, is also the world in which humans act politically. *Homo faber*, in this sense, prepares the stage for our political self, our *zoon politikon*, to appear.

This close relationship between the world and its being a space of appearance for *praxis* shows that fabrication, namely *poiesis*, has a significant role in the possibilities of *praxis*'s appearance. Even though Arendt repeatedly says that a space becomes a *space of appearance* (namely public space) "wherever men gather in the manner of speech and deeds" (Arendt, 1998, p. 199), this does not obviate the necessity for the physical-material existence of such a space, which can turn into a space of appearance. This world that *homo faber* builds, as mentioned before, houses a plurality of appearances, in addition to housing artificial things. In that sense, it is the common reference point for the citizens to acknowledge when they are disclosing themselves and how this world appears to them. This world that we live in, the one that we love and care for, relates and separates us at the same time. By virtue of this quality, it also creates common care and common concerns for the individuals, and each individual shares their care and concerns with others. That is why, for Arendt, we are worldly beings.

There is one thing to note here: Even though Arendt talks about things, artificiality, and human constructs, these very rarely refer to a specific physical or material object or artifact, and in her general political theory, the physicality of the end product of *homo faber* is left underdeveloped. While Arendt acknowledges that *homo faber*, while carrying out *poiesis*, produces artificial things in the form of both physical or material objects and cultural or institutional bodies, she does not develop the political relevance of the material or physical objects. One of the tasks at hand in this book is to develop this mute aspect of Arendt's political theory. In the subsection below, I seek to find hints from her work to provide the reader with insight into artificiality as Arendt perceives it, and I analyze if and how these hints could be referring to materiality.

The World

Arendt's political theory revitalizes the public quality of politics, aiming to situate human existence with an ontological connection to the world, a connection that can only be meaningful when it is revealed in the manner of politics. In other words, the world that is the accumulation of the end products of *homo faber* only becomes meaningful when it is appropriated by *zoon politikon*. Arendt uses the terms public space, public realm, the world, and the space of appearance interchangeably. This may confuse the reader. Therefore, I shall take a moment here to provide some distinctions between these terms.

The earth that humans inhabit is natural; the world is what humans create by intervening into the earth. The world Arendt has in mind, as mentioned before, is an accumulation of the creations of *homo faber*. In that sense, the world is separate from the earth. *Homo faber*'s creations are of two types: The first includes physical, material creations like buildings, roads, tables, and chairs, and the second includes cultural or institutional creations like governments, laws, and states. Both of these end products provide stability and durability to human beings. Physical, material products provide durability and stability in the sense that once, for example, a building has been erected, even when the architect of the building ceases to exist in the world, the building remains. Cultural or institutional products provide stability and durability in the sense that they are not bound to temporality. For example, the existence of a bill of rights or the law is not a temporal thing; it is written and recorded in history and law, and it stays in the world even when the writers of the bill of rights leave the world.

This world also includes two realms of existence, the public realm and the private realm. A space of appearance is a public space. The public realm of existence includes this public space, which is a temporal, abstract space that comes into being when the end product of *homo faber* is appropriated by *zoon politikon* via *praxis*. She adds:

> Everything that appears in public can be seen and heard by everybody and has the widest possible publicity. For us, appearance—something that is being seen and heard by others as well as ourselves—constitutes reality. Compared with the reality which comes from being seen and heard, even the greatest forces of intimate life—the passions of hearth, the thoughts of the mind, the delights of the senses—lead an uncertain, shadowy kind of existence unless and until they are transformed, de-privatized and de-individualized, as it were, into a shape to fit them for public appearance ... The presence of others who see what we see and hear what we hear assures us of the reality of the world and ourselves.
> (Arendt, 1998, p. 50)

The space of appearance thus comes into being when *zoon politikon* uses a space for self- and world disclosing. When *zoon politikon* appears in any space to use it for this purpose, whether that space was designated as public or private, it can turn into a space of appearance. When Arendt talks about the space of appearance, which appears at the moment *zoon politikon* appropriates the end product of *homo faber*, she discusses both physical, material end products and cultural, institutional end products. The world, being the accumulation of these products, provides room for interpretation, in Arendt's terminology, of how the world appears to one. This interpretation, as already discussed, is not possible without reference to common sense. Two people looking at the same chair may have the same or different perceptions of the chair's color. Two people looking at a law may have different interpretations of what that law entails. Our interpretations of the end products (the world) are temporal, but the end products (the world) themselves are not.

The world, because it houses the potential spaces of appearance, creates common concern and care in the people who live in it. We care about the end products of *homo faber* because they provide room for us to appear and disclose how we interpret this world that we live in—that is, to disclose how it appears to us. By interpreting the world, we disclose the world as politically meaningful. So, when Arendt talks about the care of the world, or the love of the world, she talks about the shared space that can potentially turn into a space of appearance. In this space, we can meet, discuss, and debate matters that concern us with our fellow citizens. This is how we give meaning to our lives and to the world that we live in.

As already mentioned, Arendt, looking back at the ancient Greek city-state experience, believes there to be two realms of existence in this world: public and private. Public realms are common spaces that we bring into existence when we manifest political freedom and equality (Arendt, 1998). Private realms, by contrast, are spaces in which we remove ourselves from society and focus on our own needs and survival. Below, I provide extensive definitions of these two realms of existence.

a) The Public Realm

The public realm is inclusive of spaces that are shared with other human beings. This does not mean that every space that exists within the public realm is a public space. Public spaces are common grounds for politics to appear. Public spaces are the end products of *homo faber*, as mentioned above, and are preparatory for the appearance of *praxis*. Moreover, not all public spaces are spaces of appearance. From a contemporary point of view, a public park or a city square is a public space, but this does not mean that such a space is a space of appearance.

Arendt's understanding of politics is strongly tied to public space: The coming together of separate individuals with various perspectives on how to be and how to live requires a public space. A public park, a government building, or a parliament could function as a public space. Arendt's definition of public spaces, although strongly tied to materiality, requires that such spaces have an additional character: To fulfill a political function, a public space must also be a "space of appearance" (Arendt, 1998, p. 199).

In a public space as a space of appearance, humans appear to one another through their speech and action. The space of appearance allows humans to be both plural and individual, and in addition, talking to one another and acting reveals what is real: For Arendt, politics is ontology. Self-disclosure and consensus in a public space also establish the world in a certain way and enable people to share and comprehend that world. If a person had no connection to the perceptions and experiences of others, that person would not be able to relate to other people in any way and would lose their grasp of reality.

This idea of the world being dependent on interaction in a public space can be further illustrated through the parallel that Arendt draws between performing arts and politics:

> The performing arts … have indeed strong affinity with politics. Performing artists—dancers, play-actors, musicians, and the like—need an audience to show their virtuosity, just as acting men need the presence of others before whom they can appear; both need a publicly organized space for their "work" and both depend upon others for the performance itself. Such a space of appearances is not to be taken for granted wherever men live together in a community. The Greek *polis* once was precisely that "form of government" which provided men with a space of appearances where they could act, with a kind of theatre where freedom could appear.
>
> (Arendt, 2006a, p. 152)

This quotation reveals that who and what we are, as well as what we consider real, are shaped through our public actions and speeches and through our interactions with one another. However, these processes of self-disclosure and world-disclosure can only manifest themselves in a public space, which, being a place between humans instead of belonging to and privately owned by a particular human, is durable and stable. In public spaces, stories are told and exchanged, and new stories and experiences are created. Whereas everything can vanish over time, public space, by being durable and stable, preserves humans' stories, deeds, and words through generations (Arendt, 1998). The parallel that Arendt draws between performing arts and politics also illustrates that public spaces are not passive gathering spots but rather artifacts that must be created, and creating these artifacts is a political challenge of utmost importance. Perceiving reality in its full sense and contributing to the enrichment of reality are impossible without public spaces. Identifying public spaces as artifacts also highlights their material character.

A space, in whatever form, can be claimed to be a space of appearance, and thus a public space, only if there is *praxis* being carried out in that space. What must be carefully noted here is that a space of appearance is always a public space for Arendt. All other spaces are potential public spaces within the public realm. For Arendt, the peculiarity of a space of appearance

> is that, unlike the spaces which are the work of our hands, it does not survive the actuality of the movement which brought it into being, but disappears not only with the dispersal of men—as in the case of great catastrophes when the body politic of a people is destroyed—but with the disappearance or arrest of the activities themselves. Wherever people gather together, it [the space of appearance] is potentially there, but only potentially, not necessarily and not forever.
>
> (Arendt, 1998, p. 199)

Here, Arendt wants to highlight that a space of appearance, and thus a public space, is a temporal notion. The physicality of a space is stable and durable, but the moment in which that space turns into a public space—in other words, a space of appearance—is temporal. Her main example of the public space, the ancient Greek city-state *polis*, is also not bound to physicality:

> The *polis*, properly speaking, is not the city-state in its physical location; it is the organization of the people as it arises out of acting and speaking together, and its true space lies between people living together for this purpose, no matter where they happen to be. "Wherever you go, you will be a *polis*."
>
> (Arendt, 1998, p. 198)

What the above quotation highlights is that *praxis* is a temporal activity, and its capacity to turn a place into a space of appearance, and thus a public space, is also temporary. However, it is the world, the accumulation of the end result of *homo faber*'s work, that houses this potentiality by providing a physical and cultural durability and stability. Below, I continue with the realm that Arendt finds antithetical to the public realm, the realm of the private.

b) The Private Realm

According to Arendt, the private realm of life is the realm of necessity. This realm is the house of *animal laborans*, the mode of existence where humans take care of their biological needs. This private realm corresponds to the household, because the activities that take place in this realm are mainly concerned with sustaining life. In this sphere, humans live together, but this is only for the sake of the individual; humans need the company of others to ensure their individual maintenance and survival. According to Arendt, in the ancient Greek city-state experience, all the activities that took place in the private realm were driven and ruled by necessity (Arendt, 1998). As I mentioned in Chapter 1, Arendt's notion of freedom is the freedom to move about, the freedom to exit one's privately owned place and, from that place, insert oneself into the public realm. In this sense, the private realm is the place where one can prepare for the public realm. This is the space in which one masters the biological necessities to become free. Without being free from the necessities that one takes care of in the private realm, the public realm consisting of free individuals cannot exist.

Thus, the private realm prepares humans to engage in politics. Everything that goes on in the private realm liberates humans "from the necessities of life, for the freedom of the world" (Arendt, 1998, p. 31). For Arendt, taking care of needs is a private matter, which does not have to be seen and heard by others since the activity that corresponds to taking care of needs, namely labor, is a solitary activity.

This distinction between the public and the private is established by Arendt to provide a demarcation for human activities within the world and their designated spaces. One cannot call labor a political activity, since in laboring, one neither discloses the self (i.e., establishes oneself as a distinct being) nor discloses the world (i.e., how one perceives the world). While laboring, one is only busy with one's individual needs and survival. Therefore, this activity will not be in the public eye. However, when one acts, carrying out *praxis*, one illuminates oneself as a distinct, unique being; one discloses oneself, and one also discloses the world. This disclosure has to be debated and discussed with fellow human beings to become a tangible reality. That is why *praxis* can only happen within the public realm, turning spaces into spaces of appearance.

The world in Arendt's political theory, then, is prior and preparatory to politics. What does this entail? Arendt strongly emphasizes the importance of the material, human-made world—the work of *homo faber*—that gives human beings a stable, durable place to live (Arendt, 1998). Human-made things, as Arendt argues, outlive human beings; even when we leave the world, those things remain (Arendt, 1998). Arendt herself did not explicitly connect her ideas about the material world to her political theory, keeping the realms of *homo faber* and *zoon politikon* disconnected. However, her work contains implicit connections between the two domains: The political realm of *zoon politikon* takes shape in interaction with the material world of *homo faber*. Political action requires not only a human openness to plurality but also a material space for interaction, since *zoon politikon* has to appear in some material space. Therefore, I believe there is an interdependent relationship between *zoon politikon* and *homo faber*. In fact, the character of this material space explicitly helps to shape the ways humans do politics. As mentioned before, there are two types of end products of the work of *homo faber*: the material or physical products and the cultural or institutional products. Throughout her scholarship, Arendt focused on the influence of the cultural and institutional products of *homo faber* on *zoon politikon*. However, this does not mean that she did not also consider the influence of *homo faber*'s material and physical products. Can it be argued that Arendt recognizes the influence of material and physical human artifacts on political activity? In order to answer this question below, I investigate two examples Arendt gives in *The Human Condition*.

c) Is the World a Material Construct?

In Arendt's work, examples of material spaces are the political *agora*, the *polis* with its walls, or the table around which humans gather. These places help to shape how human beings can practice politics, because they organize how humans can experience and enact plurality. From the very beginning of *The Human Condition*, Arendt discusses the interdependent relationship between things and humans:

> The impact of the world's reality upon human existence is felt and received as a conditioning force. The objectivity of the world—its object or thing character—and the human condition supplement each other because human existence is conditioned existence, it would be impossible without things, and things would be a heap of unrelated articles, a non-world, if they were not the conditioners of human existence.
>
> (Arendt, 1998, p. 9)

Throughout *The Human Condition* (Arendt, 1998), Arendt often refers to the ancient Greek *polis* as an ideal model for "political space." Interestingly enough, she implicitly connects her ideas about the relations between humans and things to her ideas about the *polis*. At first glance, Arendt seems to downplay the physical properties of the *polis*, in order to focus on the content of political interaction rather than its *context*:

> The *polis*, properly speaking, is not the city-state in its physical location; it is the organization of the people as it arises out of acting and speaking together, and its true space lies between people living together for this purpose, no matter where they happen to be … that action and speech create a space between the participants which can find its proper location almost any time and anywhere.
>
> (Arendt, 1998, p. 198)

Yet at the same time, she recognizes that this "space of appearance" always has a physical dimension:

> Action … is as though the wall of the *polis* and the boundaries of the law were drawn around an already existing public space which, however, without such stabilizing protection could not endure, could not survive the moment of action and speech itself.
>
> (Arendt, 1998, p. 198)

A public space requires "stabilizing protection" to "endure," in Arendt's view. The *polis* is a physical space of appearance where humans gather and interact, and this space, in order to be a public space, needs a demarcation: "the organization of *polis* is secured by the wall around the city" (Arendt, 1998, p. 198). This demarcation distinguishes the public from the private sphere. The political life of *zoon politikon* requires the material infrastructure of the *polis*, created by *homo faber*. The walls of the *polis* thus organize the relations between human beings in such a way that a space for political interaction is created. In addition to walls, Arendt implicitly discusses a second material artifact that organizes the content and the context of political action: the table (Arendt, 1998). She argues that living together in the world means that there is a world of things "between" human beings who have the world in common: Things connect human beings to each other in specific ways (Arendt, 1998). She writes:

> To live together in the world means essentially that a world of things is between those who have it in common, as a table located between those who sit around it, the world like every in-between relates and separates men at the same time.
>
> (Arendt, 1998, p. 52)

Material things relate human beings to each other, as the example of a table illustrates:

> The public realm, as the common world, gathers us together and yet prevents our falling over each other so to speak. What makes mass society so difficult to bear is not the number of people involved, or at least not primarily, but the fact that the world between them lost its power to gather them together, to relate and to separate them. The weirdness of this situation resembles a spiritualistic séance where a number of people gathered around a table might suddenly, through some magic trick, see the table vanish from their midst, so that two persons sitting opposite each other were no longer separated but also would be entirely unrelated to each other by anything tangible.
>
> (Arendt, 1998, p. 52)

If there were nothing tangible between individual human beings, the relations between them that make *praxis* possible would vanish. Material things organize relations of interaction. In the case of tables, it is even possible to study in detail what kinds of interaction specific table designs make possible. A square table organizes the interaction between people in a different way than a rectangular or a round table: The shape of the table helps to shape the distance between people, their orientation toward each other, and even the absence or presence of hierarchy in the interaction. Tables mediate the ways in which people present themselves, behave, and interact. Arendt even uses a concept, interest, that connects "mediation" and "politics":

> Action and speech go on between men, as they are directed toward them, and they retain their agent-revealing capacity even if their content is exclusively "objective," concerned with the matters of the world of things in which men move, which physically lies between them and out of which arise their specific, objective, worldly interests. These interests constitute, in the word's most literal significance, something which *interest,* which lies between people and therefore can relate and bind them together. Most action and speech is concerned with this in-between, which varies with each group of people, so that most words and deeds are *about* some worldly objective reality in addition to being a disclosure of the acting and speaking agent.
>
> (Arendt, 1998, p. 182)

The word "inter-est" literally means "what is in-between." If politics is about interest, this notion should also include the material objects that are "between" people and that help to shape interpersonal relations. Worldly things, like tables, chairs, or mugs, can influence the content of *praxis* and speech because they relate human beings to one another. A table creates relations between people, not only by connecting them but also by separating them. The table becomes a thing in-between. Arendt, however, uses these material artifacts metaphorically. She uses the wall of the *polis*, the physical and material end product of *homo faber*, as a metaphor for a cultural and institutional end product, the law. The table, in the example, is used as a metaphor for a matter of concern. When we care about the world that we share with others, there are matters that concern the sustainability or the maintenance of the world. These concerns are what relate us, since we care about the same matter, and these concerns are also what separate us, because we all have our individual perspectives on the matter.

The answer to the question, "Is the world that Arendt presents in her political theory a material construct?" is yes. The world consists of two types of end products of *homo faber*, but Arendt focuses on only one of these—the cultural or institutional products—in considering the ability to provide stability and durability to humans and to be the "in-between." This does not nullify the claim that *homo faber* provides the potential space for *praxis* to appear. However, it is not possible to find evidence in Arendt's work of *homo faber*'s material, physical aspect of the end product having influence on the potential of *praxis* to appear.

Arendt's focus is *praxis*, the agent-revealing and world-revealing capacity every human carries within themself. It is only through *praxis* that the world can be said to be disclosed as politically meaningful. Only our *zoon politikon* mode of being can carry this activity out. Only through *praxis* can our human life, our human experience, leave a trace on this world, because only then can we both appropriate the world as meaningful and establish ourselves as authentic beings. That is why it is important for Arendt that the hierarchy and world-disclosing capacities of different activities are maintained and that these activities are not confused with one another. Each of the activities discloses the world in a different way. Each of them is preparatory to authentic human existence, when each human as a free agent engages in political discourse among a diversity of perspectives. In the first chapter, I showed that Plato, by replacing *praxis* with *poiesis*, caused politics to be treated in terms of "in order to" relations. The problem occurs, for Arendt, when the world as it appears to *homo faber* becomes the only way in which the world can be disclosed. As has been mentioned before, Arendt has no quarrel with utilitarian and instrumental thinking about the world, since *homo faber*-ness and the activity of fabrication necessitate that form of disposition. However, this form of appearance of the world is preparatory and *a priori* for the world to appear in the manner of politics—that is, to disclose itself as politically meaningful. Therefore, *poiesis*—that is, fabrication or work itself—does not have a political quality per se, but it is closely related to politics because *poiesis* is the architect of the stage for *praxis*. In the upcoming section, I show that Arendt, while celebrating the human capacity for fabrication and artificiality, is critical of its result: modern technologies.

Technology in Arendt's Work

As mentioned, Arendt has no quarrel with fabrication (*poiesis*) and acknowledges material artifice as valuable in world making. In that sense, the material end results of the work of *homo faber* (such as roads and buildings) and the cultural end results of *homo faber* (like stories and artworks) are celebrated by Arendt. The problem, for Arendt, is modern technology. In this section, I argue that Arendt's criticism of modern technology stems from her observation that it changes the hierarchy among the activities that make up *vita activa*, causing labor to be the highest activity, rather than action. I show that the end result of this change within *vita activa* is that humans become alienated from the world to which they once were strongly attached, and they become concerned more about their own needs. Finally, because of the reasons outlined above, Arendt believes that modern technology enables the creation of mass society and thus leads to the creation of the social realm, which is neither private nor public (Arendt,1998).

Canovan (1994) argues that the artificial character of the human capacity of fabrication always has a positive connotation for Arendt. Acknowledging the truth of this observation does not, however, contradict the fact that Arendt is strongly critical of modern scientific and technological developments (Canovan, 1994). According to Canovan, this strong criticism stems from Arendt's conviction that the modern artificial world has lost its true value of durability and its fitness to house mankind (Canovan, 1994). Arendt's scholarship specifically addresses only a handful of technologies, for example, the telescope, Sputnik, flying technologies, and the atomic bomb (Arendt, 1998). While the individual technologies she assesses in her work are diverse, the vantage point of her assessment is the same.

The Human Condition opens with sharp criticism of the modern technological developments that Arendt observed in her time. In the book's prologue, Arendt analyzes Sputnik, which she describes as "an earth-bound object" (Arendt, 1998, p. 1). She continues:

> This event, second in importance to no other, not even to the splitting of the atom, would have been greeted with unmitigated joy if it had not been for the uncomfortable military and political circumstances attending it. But, curiously enough, this joy was not triumphal; it was not pride or awe at the tremendousness of human power and mastery which rilled the hearts of men, who now, when they looked up from the earth toward the skies, could behold there a thing of their own making. The immediate reaction, expressed on the spur of the moment, was relief about the first "step toward escape from men's imprisonment to the earth."
>
> (Arendt, 1998, p. 1)

Arendt does not see as negative the fact that human beings were able to send a satellite into outer space; rather, this event was one of the most important events of the twentieth century, since for the first time, mankind was able to look at the "earth from a cosmic perspective" (Arendt, 1998, p. XV). Arendt's discontent is highlighted by her quotation from one of Russia's great scientists: the launch represented "escape from men's imprisonment to the earth" (Arendt, 1998, p. 1). Arendt sees this will to escape the earth as the result of rapid scientific advancement, during which science accomplished things that humans had not been capable of imagining, and human beings not only had to rapidly adjust to those developments but were also outpaced by them (Arendt, 1998). This quotation, for Arendt, is representative of the mindset of her fellow humans at the time, which she found extremely problematic. To understand Arendt's discontent with human enthusiasm at the promise of an escape from earth-bound imprisonment, it is necessary to first understand her distinction between the earth and the world.

The Earth of Animal Laborans and the World of *Homo Faber*

Canovan argues that Arendt points to a world–earth distinction in order to discern two realms, one representing the "home that men have made for themselves" to live in, and the other representing "the natural environment which they belong to" as biological creatures (Canovan, 1994, p. 106). For Arendt, there are two aspects of human beings. On the one hand, we are biological creatures with biological needs like other animal species (Canovan, 1994, p. 106). Humans, like all other animals, are a part of nature and bound to its cyclical structure: we are born, we live, and we die (Canovan, 1994). Yet human beings also intervene into nature and construct a "world of their own over and above the natural earth" (Canovan, 1994, p. 106). Humans live on earth, and they erect a world from it by intervening in it. Thus, the world, which humans constructed themselves, is not the natural earth.

In the previous chapter, I clarified the distinctions Arendt makes between different human activities and their corresponding conditions of human life. The distinction between the earth and the world Arendt makes also corresponds to the distinction she makes between the human condition of labor and the human condition of work.

Labor, as already mentioned, corresponds to the cyclical nature of human existence: in life, a person wakes up, eats, sleeps, and does the same again and again until their life ends. In short, labor corresponds to the life process itself, and it is concerned with consumption to sustain life. According to Arendt, the survival of humans depends on their animal quality of sustaining life. Arendt says that life

> is a process that everywhere uses up durability, wears it down, makes it disappear, until eventually dead matter, the result of small, single, cyclical, life processes, returns into the over-all gigantic circle of nature herself, where no beginning and no end exist and where all natural things swing in changeless, deathless repetition.
>
> (Arendt, 1998, p. 96)

Nature, the source of everything, is the main source of labor and consumption for Arendt. However, human life is not only about survival, and it is not only experienced in the mode of *animal laborans*. According to Arendt,

> If nature and the earth generally constitute the condition of human *life*, then the world and the things of the world constitute the condition under which this specifically human life can be at home on earth … The world, the man-made home erected on earth and made of the material which earthly nature delivers into human hands, consists not of things that are consumed but of things that are used.
>
> (Arendt, 1998, pp. 135–6)

Arendt portrays the distinction she makes between the earth and the world clearly: The earth is what is natural, and the world is the product of human hands; the earth consists of things to be consumed, and the world consists of things that are used. As mentioned before, it is only through work, the human activity that corresponds to the human condition of worldliness, that the natural earth becomes "the world." While doing the activity of work, humans are *homo faber*, fabricating a variety of things, and these things constitute the human artifice (Arendt, 1998, p. 136). When humans are *homo faber*, they have certain attributes that *homo faber*-ness necessitates. According to Arendt, these attributes are instrumentalization, reliance on tools and productivity, thinking only in terms of means–ends categorization, conviction that everything can be solved, reduction of every human motivation to utility, having a sovereign attitude that views all nature as materials and nature as something to be manipulated and exhausted, equating intelligence with ingenuity, and discontent with all thought that doesn't serve a purpose for fabrication—that is, creating a tool to create a tool indefinitely (Arendt, 1998). Furthermore, Arendt argues that a feature inherent in work is reification, which she describes as manipulating and destroying whenever necessary to carry out the activity. This interference with nature has always taken the form of destruction when humans are in the mode of *homo faber*, according to Arendt (Arendt, 1998)

When we humans are carrying out the activity of work, this is how our existence takes shape, and it is actually necessary that we exist in this manner. The activity of work, fabrication, or making necessitates instrumental, utilitarian thinking; thinking in terms of means–ends categorization; and all the other attributes of *homo faber*. Work, for Arendt, entails a "utilitarian mentality, an inability to think and judge a thing apart from its function or utility" (Arendt, 2006a, p.

212). This utility, according to Villa, is a mode of understanding the world, and it is in the nature of *homo faber* to employ this mode of understanding (Villa, 1995, p. 137). According to Villa, Arendt observes in *homo faber* a pervasive "ontological pre-comprehension" (Villa, 1995, p. 138) toward the world. Because of that precomprehension, the world can only open itself up to *homo faber* in terms of utility; every possible meaning of every possible appearance is grasped within the relation of "in order to" (Villa, 1995, p. 138). One has this disposition as *homo faber* because then one is carrying out work or fabrication. Fabrication is realized with the guidance of a model or image, which does not disappear with the end product, but stays intact for future processes of fabrication (Arendt, 1998). The activity of work itself has both a definite beginning and a predictable end, and for Arendt, this is the most important feature of fabrication (Arendt, 1998). Because of this predictable end, work is reliable, and it is reversible, since humans also have the capacity to destroy everything that they produced (Arendt, 1998).

According to Arendt, because *homo faber* creates durable, tangible things that can be destroyed but are meant to stay, even after the death of their creator, this gives these durable things an objective character: Once made, they assume relative independence from their creator (Arendt, 1998, p. 137). With its durability, the human artifice interrupts the cyclical movement of nature. When we "work," we create something out of what exists in nature, and we interrupt its cyclical movement, leaving something durable and stable behind. These objective and durable creations of work constitute the human-made world and function for "stabilizing the human life" (Arendt, 1998, p. 137). For Arendt, the objectivity of these creations "lies in the fact that ... men, their ever-changing nature notwithstanding, can retrieve their sameness, that is their identity, by being related to the same chair and same table" (Arendt, 1998, p. 137). The world's objectivity provides a realm for people to express their subjectivity by having the same kind of bond with the same thing. In other words, adds Arendt,

> against the subjectivity of men stands the objectivity of the man-made world rather than the sublime indifference of an untouched nature, whose overwhelming elementary force, on the contrary, will compel them to swing relentlessly in the circle of their own biological movement, which fits so closely into the over-all cyclical movement of nature's household.
>
> (Arendt, 1998, p. 137)

The world protects humans from the cyclical movement of nature by being something permanent and allows them to look upon nature as something "objective" (Arendt, 1998, p. 137). If there were no permanent place for humans to express their subjectivity and to be protected from nature, there would be no space for them to create a life story of their own for others to hear after their death. Without a world between humans and nature, humans would simply exit this world as biological creatures, would have no distinct traits attributed to them, and would not be able to leave a trace behind. Arendt says: "without being at home in the midst of things whose durability makes them fit for use and for erecting a world whose very permanence stands in direct contrast to life, this life would never be human" (Arendt, 1998, p. 135).

In sum, by making a distinction between the earth and the world, between the dispositions of *animal laborans* and *homo faber*, Arendt points out the biological conditions we share with all the species of the earth and the artificial conditions we create to distinguish ourselves. For Arendt, both the world and the earth are important for human life. The scientific and technological advancements she observed in her time, though, were challenging the distinctions she presented. Scientists taking the first "step toward escape from men's imprisonment to the earth" (Arendt, 1998, p. 1) point to a very strong will that can annihilate all human life, according to Arendt.

Even though Arendt celebrates the artificial human life that work represents, a truly artificial life would, she argues, cut humans' ties to their humanness. A completely artificial life is a major sign of danger for Arendt, since she believes such a development event would cause a dramatic change in the way in which we live.

As mentioned before, Arendt values both *animal laborans* and *homo faber* in their corresponding activities: labor and work. The problem, for Arendt, emerges when these activities are intermingled with one another without a clear distinction between them. It is important to remind the reader that Arendt positions these activities relative to human dispositions toward the world. Each disposition is connected with an activity, a perspective toward the world, and a realm where it is appropriate for this activity to take place. This is the reason behind the hierarchy of activities within *vita activa*. The advent of modern technology, automation, and machinery, and the accompanying new mindset of humans, caused a change in the hierarchy of the activities within *vita activa*, according to Arendt. She calls the end result of the change in the hierarchy of the activities within the *vita activa* world alienation.

World Alienation

In *The Human Condition* (Arendt, 1998), Arendt claims—contrary to Marx—that it is world alienation, "not self-alienation, that is the hallmark of the modern age"[5] (Arendt, 1998, p. 254). World alienation, for Arendt, is the loss of what is common to all humans, and modernity is the cause of world alienation. Modern technological and scientific developments have transformed the ontological question that humans ask about the world, from "What is this?" into "What is this for?" (Tijmes, 1992). As discussed above, this is the mentality of *homo faber*. The more technological and scientific advancements humankind has had, according to Arendt, the more humans have become concerned with the latter question. Tijmes (1992, p. 392) also argues that Arendt's concept of world alienation is mainly about humans' detachment from their artificial inheritance, the world (Tijmes,1992, p. 392). The accumulation of the end products of *homo faber*, the world, has lost its capacity to be the "in-between" that relates and separates us, because human experiences "have been reduced to subjective, private ones" (Tijmes, 1992, p. 392). With the beginning of the modern age, Arendt believes the hierarchy of the activities within *vita activa* started undergoing a radical upheaval. Action, which was the highest form of activity, first lost its rank to work (fabrication), and this loss was followed by labor assuming the highest rank among the activities.

Arendt starts to unravel this reversal within *vita activa* by explaining the victory of *homo faber*, since for her, *homo faber* was the first to triumph in this upheaval. The work of *homo faber* was valuable because its expertise—the creativity and talent to turn what is natural into objects—was essential to trap nature and make experiments (Canovan, 1994). The problem here, according to Arendt, is not that *homo faber* creates worldly objects—since that is what *homo faber* does to provide a stable and durable environment for humans—but rather that there is a new worldview in which *homo faber* repeats this creation of objects endlessly where there is no use for them. The concern of *homo faber* stops being about the ideas, models, or shapes to come, and instead is about the process in which these artifacts are realized (Arendt, 1998).

As mentioned before, *homo faber*, according to Arendt, is utilitarian and manipulates nature, producing and creating materials from it, and the value of these products could only be measured in terms of their utility. However, in the new worldview, this concept of valuing things according to their utility was transferred to valuing the process that brings these objects into being, and this mindset became the perspective ruling how one perceives the world itself.

If, Arendt continues, these worldly objects are not valued in terms of their usefulness but as the unintended results of a production process, then the end result of the production process (the product itself) loses all its meaning (Arendt, 1998). This leads to a new definition of *homo faber* as the builder of human artifice who accidentally invents tools (without having the intention of creating them), rather than being *homo faber*, the maker of objects. As Tijmes argues, for Arendt, the world that we create as a result of our work, in our modern times, no longer represents for us "the touchstone of reality" (Tijmes, 1992, p. 392).

If this principle of the utility of *homo faber* is applied not to the worldly objects but to the production process, the definition of what is useful shifts toward things that stimulate productivity and decrease pain and effort. This creates a shift in our understanding of the world as well. *Homo faber* becomes fulfilled by the endeavor of creating for the sake of creating, rather than creating for something, creating "in order to." Arendt says,

> Only in a strictly anthropocentric world, where the user, that is, man himself, becomes the ultimate end which puts a stop to the unending chain of ends and means, can utility as such acquire the dignity of meaningfulness. Yet the tragedy is that in the moment *homo faber* seems to have found fulfilment in terms of his own activity, he begins to degrade the world of things, the end and end product of his own mind and hands; if man the user is the highest end, "the measure of all things," then not only nature, treated by *homo faber* as the almost "worthless material" upon which to work, but the "valuable" things themselves have become mere means, losing thereby their own intrinsic "value."
>
> (Arendt, 1998, p. 155)

Thus, *homo faber*, starts considering human beings as the measure of all things, rather than maintaining a holistic perspective of the world as a whole. The standard measurement shifts from utility, in the sense of being useful for something, to happiness, "that is, the amount of pain and pleasure experienced in the production or consumption of things" (Arendt, 1998, p. 309). This leads to a translation of *homo faber*'s values into the values of *animal laborans*: The subjective experiences, the pain and pleasure of humans, what humans want only for themselves and what they need "becomes the highest good" (Canovan, 1994, p. 153).

Where, then, does this bring the discussion to? The victory of *animal laborans* brought about the last stage of the laboring society, the society of jobholders

(Canovan, 1994). The problem with the victory of *animal laborans*, according to Arendt, is that labor is an anti-political activity. While laboring, humans are neither with the world nor with one another but alone with their own bodies, taking care of necessities. *Animal laborans* is, of course, living in a world where others are present, but one cannot call this human plurality; it is rather like being in the company of friends, enjoying leisure time together (Arendt, 1998). Laboring is the same for all, and a society where everybody labors and consumes leads to an "experience of laboring together, where the biological rhythm of labor unites the group of laborers to the point that each may feel that he is no longer an individual but actually one with all others" (Arendt, 1998, p. 214). In the society of *animal laborans*, all individuality is submerged in the "over-all life process of the species" (Canovan, 1994, p. 153). With the victory of *animal laborans*, where there is no plurality but sameness, Arendt's biggest fear, mass society, comes into existence.

I have outlined that Arendt believes world alienation to be the end result of modern technological artifice. I do not want to claim that Arendt was against any technological progress, but she was obviously discontented with the mindset that enabled the technological developments she observed in her time. To Arendt, the problem is not only the technological products of her time but the fact that these products no longer have the worldly capacity of human artifice. They are created to make the life of *animal laborans* easy, to make *animal laborans* consume, and to keep humans in the mode of *animal laborans* so that they do not act within the world. The more the needs of *animal laborans* are satisfied, the more humans will lose their care for the world and replace it with care for the self. This, for Arendt, is worldlessness, which brings with it a realm of human life that blurs all the distinctions between human activities and their corresponding realms.

Worldlessness and the Social Realm

According to Arendt, when *animal laborans* assumes the highest rank within the activities of *vita activa*, humans become worldless. The worldlessness of *animal laborans*, Arendt says, is a condition in which humans are "imprisoned in the privacy of their body" (Arendt, 1998, p. 118), all activities being subsumed to taking care of their needs. When humans are in the mode of *animal laborans*, their activities remain private; they do not need the existence of others to do what is necessary to take care of their own needs. But the mode of *animal laborans*, or carrying out labor, cannot be our only mode of existence. In a society of *animal laborans*, the world stops being a realm that relates and separates us from others, since we are deprived of the "intermediary of a common world of things" (Arendt, 1998, p. 58). The collective existence of these individuals leading private lives, and thus living only within the mode of *animal laborans*, is described by Arendt as mass society. This mass society exists in what Arendt calls "the social realm" or simply "the social" (Arendt, 1998, pp. 38–40, 45, 209, 257).

The social realm is the end result of private matters making room for themselves in the political arena, with the result that people's talk becomes entirely concerned with such private matters. According to Arendt, society "always demands that its members act as though they were members of one enormous family which has only one opinion and one demand" (Arendt, 1998, p. 39). This kind of behavior, which Arendt calls conformity of opinions and demands, was present in the family, which was in the realm of the private. With the rise of the social, appearing as a distinct, unique human being became a private matter. Thus, the driving motive of action, which is realizing one's own self by appearing in public with public speaking of words and public performance of deeds, has been replaced by private interests, which are temporary and limited to the life of the human being.

As already indicated, for Arendt, the rise of the social was fostered by the modern industrial revolution (Arendt, 1998). The entrance of economic concerns into the public realm inevitably overwhelmed the matters people were concerned with. In this sense, Pitkin (1998) argues that Arendt's understanding of

> the social realm seems to involve the development of a complex economy: trade, money, division of labor, a market system, and eventually the extensive, centralized, intricate network of production and exchange of our time in which people are profoundly interdependent, yet no one is in charge, outcomes being determined, as we say, by "the market."[6]
>
> (Pitkin, 1998, p. 11)

Certain groups of people lost their privately owned place in the world, and this situation exposed people to the most urgent needs of life, namely labor (Arendt, 1998). When this labor was discovered to be something that can also be turned into an accumulation of wealth and capital, this caused the inevitable rise of capitalism (Arendt, 1998). The laboring class, since their whole existence was reduced to the worries of immediate needs, became alienated from other worries, such as care for the world (Arendt, 1998). For Arendt, it was Max Weber who made the most accurate discovery of the origins of capitalism, demonstrating that "an enormous, strictly mundane activity is possible without any care for or enjoyment of the world whatever, an activity whose deepest motivation, on the contrary, is worry and care about the self" (Arendt, 1998, p. 254). However, this is not the only feature of the social realm. Aside from bringing private concerns to public light, the social also creates a new form of living together in the world, and Arendt calls this form "society" or "mass society" (Pitkin, 1998). Mass society[7] consists of individuals who are only concerned about their own life processes, and the world appears as a realm in which people behave the same, need the same things, and display their private concerns out in the open. The public realm is no longer a space of appearance, being reduced to a space of sameness.

Society, for Arendt, is the accumulation of private individuals concerned with their own needs. The private individuals in society do not coexist, although they are dependent on one another to be able to pursue their individual goals and aims. This codependence, rather than coexistence of human beings, is what makes Arendt believe humans are losing the room to act politically, and their disposition toward the world manifests itself in the form of *taking care of needs*.[8] If a concern does not refer to taking care of some form of need, it is not relevant for society. This category of needs, according to Arendt, has a wide spectrum that can cover almost all forms of public services, such as providing health care and maintenance of public transportation, or basic human needs such as food and shelter.

Adequate housing, for example, according to Arendt, is not a political matter but rather a social matter, an obvious need that should be fulfilled: Every human being should have the possibility to put a roof over their head, and there is no debate to be had about this.[9] Moreover, needs are "coercive" in nature (Schaap, 2010). They refer to inclinations directed at survival and, therefore, do not belong to the realm of the political[10] (Schaap, 2010). Arendt's conception of "needs" can be further clarified by her reading of the 1789 French Revolution, which she sees as driven by a necessity, namely eradicating poverty. She argues:

> Poverty is more than deprivation, it is a state of constant want and acute misery whose ignominy consists in its dehumanizing force; poverty is abject because it puts men under the absolute dictate of their bodies, that is, under the absolute dictate of necessity ... when they (the poor) appeared on the scene of politics, necessity appeared with them, and the result was that ... freedom had to be surrendered to necessity, to the urgency of life process itself.
>
> (Arendt, 2006b, p. 50)

What this quotation shows is that, once our physical needs are not met due to poverty, we cannot be driven by anything but the dictates of satisfying those needs. This dictate does not allow us to develop other and different dispositions toward the world beyond our pursuit of food; the pursuit of food becomes our only disposition. All our actions and the way we think become dependent on this need. Moreover, for Arendt, the French Revolution did not overcome the "need disposition," which is also reflected in the focus on "need" in the public realm in our current situation in politics today. Although the poor became wealthy, public debates have not progressed beyond the "motive of need." The oppressed

> did not become men of leisure whose actions were prompted by a desire to excel, but succumbed to the boredom of vacant time, and while they too developed a taste for "consideration and congratulation," they were content to get these "goods" as cheaply as possible, that is, they eliminated the passion for distinction and excellence that can exert itself only in the broad daylight of public. The end of government remained for them self-preservation that "it is a principal end of government to regulate (the passion for distinction)" has not even become a matter of controversy, it is simply forgotten. Instead of entering the marketplace, where excellence can shine, they preferred, as it were, to throw open their private houses in "conspicuous consumption," to display their wealth and to show what, by its very nature, is not fit to be seen by all.
>
> (Arendt, 2006b, p. 60)

What Arendt criticizes is not only that consumption itself has become a need but also that it has become a need to show others that one can consume (Arendt, 2006b). Instead of going beyond "need as basic motive," the people who overcame poverty found new and different needs, which became the sole concern of current politics in our consumer society; politics has become solving economic and financial problems—in one word: administration. Arendt's distinction between the political and the social indicates a certain order: If we want to engage in political interaction, "we first need to be liberated from nature by satisfying our biological needs for food, shelter and so on" (Schaap, 2010, p. 159). Arendt does not mean by this that people in poverty or hunger should not be allowed to enter into a public debate, but rather that fulfilling those needs should not be the subject of the debate. For Arendt, biological needs and poverty are examples of coercive conditions that cannot be transcended or overcome.

Her main aim in distinguishing the social from the political is to demarcate a political realm that is not dictated by needs but in which there is freedom from needs. According to Schaap (2010), for Arendt, any social movement or collective organization that concerns itself with satisfying needs should be considered as nonpolitical (Schaap, 2010). For Schaap, this also implies that questions of social justice are not political issues, since they too belong to this category of needs (Schaap, 2010). A further consequence of the rise of the social realm, for Arendt, is that "our capacity for action and speech has lost much of its former quality since the rise of the social realm banished these into the sphere of the intimate and the private" (Arendt, 1998, p. 49). Thus, according to Arendt, in the turn to modernity, the political (which she reserves for the human capacity for action and speech) is absorbed by the social.

It is clear that Arendt finds modern artifice, and the accompanying mindset that enables it, problematic. For Arendt, the technological advances of the modern age bring with them what she calls "world alienation," which she defines as humans' estrangement from the world they live in. This, according to Arendt, is caused by producing things rather than erecting a world. With modern artifice, Arendt claims that the material objects are there only to be consumed and discarded.

When one reads Arendt's equation of modern scientific and technological advancements with world alienation, it becomes hard to separate her views from those of classical philosophers of technology. According to Verbeek (2005), these classical philosophers present a standard picture of technology (Verbeek, 2005, p. 2) that depicts it

> as a radically transformative power that estranges human beings from themselves, from each other, and from reality itself. Technological culture is seen as transforming human beings into cogs in a social machine, and as transforming reality into raw material that can only be approached via domination and control.
>
> (Verbeek, 2005, p. 3)

It should not be surprising that Arendt presents modern technology in accordance with this standard picture. After all, she was the student of both Karl Jaspers (2014) and Martin Heidegger (1962), who—Verbeek claims—evidently subscribed to this standard "classical conception of technology" (Verbeek, 2005, pp. 3–10).

According to Verbeek, Jaspers argues that technology's influence on human existence is transformative in the sense that it creates large-scale "mass rule," fostering mass production and mass culture while promoting "an entirely new way of existence in an entirely new material environment" (Verbeek, 2005,

p. 10). Like Arendt, Jaspers perceives technology as bringing about alienation for humans, both from themselves and from the world (Verbeek, 2005, p. 10). For Jaspers, this is the end of authenticity, and to overcome this loss of authenticity, humans need to realize the instrumental nature of technologies: They are means to achieve goals that humans set for themselves (Verbeek, 2005, p. 10). The more humans perceive technology as instrumental, the less the humanness of humans will be threatened by technology. Compared to Jaspers, Heidegger is more focused on analyzing the role of technology in the relation between humans and the world (Verbeek, 2005, p. 10). In that sense, according to Verbeek, Heidegger perceives technology as a "particular manner of approaching reality" (Verbeek, 2005, p. 10) with which humans seek to dominate, control, and manipulate the world (Verbeek, 2005, p. 10). Both these approaches to technology, however, were objects of criticism. Verbeek (2005) argues that classical philosophers of technology painted a rather pessimistic picture while perceiving technology "with a capital *T*, leaving no room for different kinds of descriptions of different kinds of technologies" (Verbeek, 2005, p. 4). Verbeek here claims that, rather than focusing on individual technologies and trying to understand their specific applications (Verbeek, 2005, p. 5), classical philosophers of technology perceived technology in terms of "what must be presupposed in order for it to be possible" (Verbeek, 2005, p. 7). Instead of focusing on what technologies are *making possible*, classical philosophers of technology adopted a backward perspective in an attempt to grasp what *made* the technologies we have *possible* (Verbeek, 2005, p. 7). According to Verbeek,

> Classical philosophy of technology prestructured its analysis in such a way that it could not *but* discover alienation. It failed to see that the diagnosis that technology *presupposes* a dominating manner of thinking, or a functional orientation of social life does not necessarily imply that dealing with concrete technologies can only *produce* this domination and functionalism.
>
> (Verbeek, 2005, pp. 7–8)

Verbeek's criticism of both Jaspers and Heidegger is also applicable to Arendt. The main problem with Arendt's analysis of modern technology is that she situates technologies within the realm of the social and its accompanying mode of worldlessness. I believe that Arendt takes a limited approach to her own insight into the work of *homo faber*. In her reading, modern technology becomes detrimental to politics. I argue that Arendt left the work of *homo faber* underdeveloped in her characterization of the world itself and began to be more concerned with the loss of the world. She herself situates worldliness as the accompanying disposition to *homo faber*. In Arendt's hierarchy of *vita activa*, *homo faber* creates artifacts so that *zoon politikon* can appear and flourish. These material objects are produced for humans to use and make sense of, to value them, and to appropriate them. The world in common allows people to join together in action and speech, but it also allows people to distinguish themselves and be recognized as distinct individuals who share the world with their fellow human beings. Modernity, for Arendt, makes the end result of work lose its meaning and value. This belief makes Arendt limit technology (an example of material artifice) to its conditions of possibility. However, I propose taking Arendt's analysis of the cultural end products of *homo faber*, like stories and artworks, as a point of departure, rather than her limited reading of technologies for the politics of technology. I argue that, once unpacked, Arendt's description of stories and artworks can guide us toward an interaction-based politics of technology.

Cultural Products of *Homo Faber*

In this section, I show that Arendt does not consider all end results of *poiesis* as irrelevant to politics. I use three steps to argue that, for Arendt, the cultural products of *homo faber* not only stabilize the appearance of *praxis* retroactively but also influence the further appearance of *praxis* proactively. First, I show that for Arendt, common sense is the accumulation of the records of the temporary appearances of *praxis*. Second, following Arendt's own examples, I show that the cultural products of *homo faber*—as in stories or theater plays—are the embodiment of the common sense of their time of creation. This, I show in the third step, proves that these cultural products of *homo faber* not only stabilize the appearance of *praxis*, by turning it into these products, but also inspire further appearances of *praxis* because these cultural products are an embodiment of the meaning that the world has been given at their time of creation.

Coproduction of Common Sense and Remembrance

As mentioned before, *praxis* almost always assumes a common reference point between acting agents, which is the world—in other words, a shared common world that relates us to and separates us from other acting beings. This common world, as mentioned, is the accumulation of the creations of *homo faber*, whether cultural and institutional or physical and material. This implies that *praxis* is dependent on the products of *homo faber* because, without the common world as a common reference point that is stable, durable, and objective, we would not be able to grasp how humans are related to and separated from one another. In Arendt's words,

> Action and speech go on between men, as they are directed toward them, and they retain their agent-revealing capacity even if their content is exclusively "objective," concerned with the matters of the world of things in which men move, which physically lies between them and out of which arise their specific, objective, worldly interests. These interests constitute, in the word's most literal significance, something which *interest,* which lies between people and therefore can relate and bind them together. Most action and speech is concerned with this in-between, which varies with each group of people, so that most words and deeds are *about* some worldly objective reality in addition to being a disclosure of the acting and speaking agent.
>
> (Arendt, 1998, p. 182)

This quotation reiterates that our common intersubjective world is what creates a sense of togetherness through relating and separating us at the same time. Arendt's words, however, also simultaneously show that what the world *means* depends on acting agents who constantly reveal how the world makes sense to them. Because of our *amor mundi,* we also want to maintain this world, not only as a space that houses our authentic appearance, as individuals and as a community, but also as a space for generations to come. As Arendt puts it, "men initially entered the public realm because they wanted something of their own or something they had in common with others to be more permanent than their earthly lives" (Arendt, 1998, p. 55). She also highlights the necessity of maintaining the world as inclusive of public spaces for future generations because "if the world is to contain a public space, it cannot be erected for one generation and planned for the living only" (Arendt, 1998, p. 55). This begs the question of how this maintenance of the public space for future generations is to be done.

Is it not common sense that discloses and at the same time maintains the world? (Borren, 2013). If we accept this, which I have been assuming until now that we do, then would it not be fair to say that it is common sense that maintains the *meaningful* appearances of *praxis* by being an accumulation of the records of these temporary appearances? Would it be too far-fetched, then, if I claim that it is *poiesis*, the work of *homo faber*, that records these temporal appearances of *praxis*? If common sense is the basis for our preliminary understanding of the world, through which we can interpret and understand what the world means for us and share that meaning with our fellow human beings, and if common sense is the accumulation of previous appearances of *praxis*, then it must be *poiesis* that enables us to have access to these previous appearances of *praxis*. Otherwise, how would we know what the world is? We have a preliminary understanding of the world because we have the records. For example, we know that the Hungarians revolted against the Soviet regime in 1954 because this temporary appearance of *praxis* was stabilized by *poiesis*: It was turned into a story for us to know, to add to our store of common sense. We know this story because we share a common world and a common sense. Arendt never described the relationship between *poiesis* and *praxis* as I am doing here, but she gave hints of it. By providing examples from Arendt herself, I will show that the products of *homo faber* not only stabilize the appearance of *praxis* retroactively but also influence the further appearance of *praxis* proactively.

The work of *homo faber*, by providing stability and durability, allows the appearance of *praxis* to be preserved and thus not forgotten. As humans, we have a limited time on earth: Like all animals, we are born into this world, and we die. What makes us distinct is our capacity for *praxis*. Through the preservation of our *praxis*, we as individuals will be remembered. Furthermore, the appearance of the world in accordance with our disclosure will also be recorded as one way, among many, in which the world appeared at the time when and in a place where freedom and plurality were worldly existences. In Arendt's words,

> The whole factual world of human affairs depends for its reality and continued existence, first, upon the presence of others who have seen and heard and will remember, and, second, on the transformation of the intangible into the tangibility of things. Without remembrance and without the reification which remembrance needs ... the living activities of action, speech, and thought would lose their reality at the end of each process and disappear as though they never had been.
>
> (Arendt, 1998, p. 95)

This quotation shows that Arendt acknowledges the temporal character of *praxis* and argues that *praxis* must be recorded and maintained within the world. Only through remembrance can *praxis* inform and inspire our future activities. When *praxis* appears, it discloses the world—that is, how the world appears to the agent carrying out *praxis*—and it discloses the agent to the audience—that is, how the agent appears to others. Through this disclosure, we show how the world is politically meaningful and how we situate ourselves within this world. However, the meaning we ascribe to the world—the meaning of our appearance and the intersubjective relationships that we engage in—would be lost if it were not remembered. I argue that it is *homo faber* who translates the appearance of *praxis* into a tangible form so that it will not be forgotten. It is these very translations made by *homo faber* that maintain and preserve the appearance of *praxis*: how the world and the agent carrying out *praxis* were disclosed. The work of *homo faber*, then,

> prepares the intangible and the futile for their eventual materialization; it is the beginning of the work process, and like the craftsman's consideration of the model which will guide his work, its most immaterial stage. The work itself then always requires some material upon which it will be performed and which through fabrication, the activity of *homo faber*, will be transformed into a worldly object.
>
> (Arendt, 1998, pp. 90–1)

The question, then, is the following: How does *homo faber* turn the intangible and futile into something tangible? In the following subsection, based on Arendt's examples, I show that *homo faber*, by remembering and by making something beautiful, turns the temporary appearance of *praxis* into stories and artworks.

Cultural Artifacts

a) Stories

Arendt argues that because *praxis* is inherently unpredictable, it is thus impossible to foresee what the consequences of its appearance could be, and people start to gather what the appearance of *praxis* meant only after the deed is finalized (Arendt, 1998). This means that people could know the meaning of the Hungarians demolishing the statue of Stalin at the edge of Varosliget Park only after the statue was toppled. The people who participated in or observed the event could recall what happened and tell others the story of the events and arising relationships in that park on October 23, 1954. The meaning of this appearance of *praxis* thus "arises directly out of the story which, as the result of action, begins and establishes itself as soon as the fleeting moment of the deed is past" (Arendt, 1998, pp. 191–2). Arendt continues:

> whatever the character and content of the subsequent story may be, whether it is played in private or public life, whether it involves many or few actors its full meaning can reveal itself only when it has ended … Action reveals itself fully only to the storyteller, that is, to the backward glance of the historian, who indeed always knows better what it was all about than the participants … Even though stories are the inevitable results of action, it is not the actor but the storyteller who perceives and "makes" the story.
>
> (Arendt, 1998, p. 192)

Here, Arendt positions the historian as the spectator who looks at an event and judges what it means. According to Borren, the spectator is the one who can reach common-sense judgments, who neither has the first-person perspective that is completely immersed in phenomena nor that of the observer who claims to have a neutral point of view (Borren, 2013). Borren identifies the notion of the spectator as Arendt's "methodological treasure" (Borren, 2013, p. 244). But the spectator is not only an observer with situated impartiality; they are the *zoon politikon* and the *homo faber* at the same time. As Arendt herself says, "the public realm is constituted by the critics and the spectators … and this critic and the spectator sits in every actor and fabricator" (Arendt, 1982, p. 63). *Homo faber* as the spectator turns *praxis* into a story through craftsmanship, thus making the intangible appearance of *praxis* into a tangible reality. The spectator who has situated impartiality, as Disch explains it, is the storyteller who "teaches her readers to see as she does, not what she does, affording them the 'intoxicating' experience of seeing from multiple perspectives but leaving them with the responsibility to undertake the critical task of interpretation for themselves" (Disch, 1993, p. 687). Or, as Arendt puts it,

> The task of the poet and historiographer (both of whom Aristotle still puts in the same category because their subject is (praxis)) consists in making something lasting out of remembrance. They do this by translating action and speech, into fabrication which eventually becomes the written word.
>
> (Arendt, 2006a, p. 44)

Arendt goes on to argue that

> acting and speaking men need the help of *homo faber* in his highest capacity, that is, the help of the artist, of poets and historiographers, of monument-builders or writers, because without them the only product of their activity, the story they enact and tell, would not survive at all. In order to be what the world is always meant to be, a home for men during their life on earth, the

> human artifice must be a place fit for action and speech, for activities not only entirely useless for the necessities of life but of an entirely different nature from the manifold activities of fabrication by which the world itself and all things in it are produced.
>
> (Arendt, 1998, p. 173)

Making a story out of the appearance of *praxis*, then, is how *poiesis* turns *praxis* into a tangible reality and thus maintains the temporal appearance of *praxis* as something durable and stable. In *On Revolution*, Arendt argues that it is not the actors themselves who can communicate the meaning of the appearance of their *praxis* but the writers or historians, who will lay bare the meaning of what the actors did. She says

> That the men of the French Revolution should have been unable to think in these terms, and therefore never really touched the heart of the matter which their own actions had brought to the fore, is actually almost a matter of course. Obviously, they knew at most the principles that inspired their acts, but hardly the meaning of the story which eventually was to result from them. Melville and Dostoevski, at any rate, even if they had not been the great writers and thinkers they actually both were, certainly were in a better position to know what it all had been about.
>
> (Arendt, 2006b, pp. 73–3)

So, in a translation of the appearance of *praxis* into a story by *poiesis*, there can be seen not only an engagement with common sense through stabilizing it into a text but also a "reifying" of how it appeared. One can hear and read the stories of the heroes who revealed the world as a space of appearance that exhibited plurality and freedom.

b) Artworks

The second form in which *poiesis* turns *praxis*'s appearance into a tangible reality is in artworks like paintings, sculptures, or even architectural works like cathedrals. According to Arendt,

> The cathedrals were built *ad maiorem gloriam Dei*; while they as buildings certainly served the needs of the community, their elaborate beauty can never be explained by these needs, which could have been served quite as well by any nondescript building. An object is cultural to the extent that it can endure; this durability is the very opposite of its functionality, which is the quality which

> makes it disappear again from the phenomenal world by being used and used up. The "thingness" of an object appears in its shape and appearance, the proper criterion of which is beauty. If we wanted to judge an object by its use value alone, and not also by its appearance (that is, by whether it is beautiful or ugly or something in between) we would first have to pluck out our eyes.
>
> (Arendt, 1960b, p. 286)

Here, Arendt presents another aspect of the end result of the work of *homo faber* and how it translates *praxis* into a worldly, tangible reality: by translating it into a beautiful work of art. *Poiesis* allows the appearance of *praxis* to be maintained by translating its appearance into stories, and it allows the appearance of *praxis* to be deemed beautiful by turning it into works of art, whether in the form of a painting or in the form of a cathedral.

One can raise here the question: Is it not then *poiesis* that actually discloses the world as politically meaningful by translating the appearance of *praxis* into stories, architecture, and art? Alternatively, if it really is *poiesis* or *homo faber* that allows *praxis* to be maintained and stabilized, is this not also problematic? After all, *homo faber* is presented as our mode of being that seeks utility, perceives the world only in terms of means–ends relationships, and the end products of whose work are instrumental. This dependence of *praxis* for remembrance on *poiesis* can lead one to conclude that what the appearance of *praxis* meant is actually dependent on the author, painter, or architect, namely on *homo faber*. How the fabricators perceived the appearance of *praxis* is inscribed in their works, and that is the only way in which we can *see*, or *read*, this appearance. This would shift the locus of politics as interaction too, since now, the meaning attributed to the appearance of *praxis* is sought not in what appears during the intersubjective relationship that happened in the world but in how *homo faber* tells what appeared.

However, it should be remembered that *homo faber*'s end products, whether cultural, like those discussed above, or material, are not fixed in terms of meaning. According to Disch (Disch, 1993), a cultural product of *homo faber*, such as a story, "can be both ambiguous and meaningful at once" (Disch, 1993, p. 688). She continues:

> An ambiguous argument, testimony, or example is less effective for its indeterminacy, because the purpose of such modes of discourse is to distill the plural meanings of an incident into definitive conclusions. Ambiguity in a

> story encourages the permanent contestation and multiple reinterpretation of meanings that make situated impartiality possible.
>
> (Disch, 1993, p. 688)

What I want to emphasize in this quotation is "permanent contestation and multiple reinterpretation" (Disch, 1993, p. 688). This is exactly why the cultural products of *homo faber*, which are *poiesis*, are also dependent on *praxis*. These products, in order to remain in the world as meaningful or beautiful and to provide the previous appearances of *praxis* with the desired stability and durability, have to be appropriated by *praxis* again and again. In Arendt's words,

> To be sure, an ordinary use object is not and should not be intended to be beautiful; yet whatever has a shape at all and is seen cannot help being either beautiful, ugly, or something in-between. Everything that is, must appear, and nothing can appear without a shape of its own; hence there is in fact no thing that does not in some way transcend its functional use, and its transcendence, its beauty or ugliness, is identical with appearing publicly and being seen.
>
> (Arendt, 1998, pp. 172–3)

So, while *praxis* is dependent on *poiesis* to be maintained and remembered, and thus to contribute to common sense to inform future potential appearances of *praxis*, *poiesis* is also dependent on *praxis* to be appropriated, maintained, and consumed.

Let me illustrate with an example. *Homo faber* can create an artwork to allow the community to see what he saw as a spectator. Let us assume this spectator wants to turn the temporal appearance of the revolutionaries during the Hungarian Revolution of 1956 into something stable, and so he creates a sculpture. After its creation, what that sculpture *means* is based on its appropriation through *praxis*, even though *homo faber* created the sculpture as a spectator first, as a testimony to the appearance of *praxis*. Yet the meaning that is going to be ascribed to that sculpture cannot be fixed by *homo faber*. Now that the sculpture is out in the world, it is an appearance that will be perceived by the acting agent, and its meaning will depend on the hermeneutic circle that will form once the agent engages with the sculpture. In other words, it is the interpretative act that prospectively creates durability, as well as the act of creation.

What needs to be highlighted here is that neither meaning nor beauty can be fixed into an artifact. These notions are agreements that we make with our fellow human beings because both notions are based on judgments that we form by referring to our common sense. When I perceive a phenomenon, I can only grasp what that phenomenon means or judge if it is beautiful based on what is disclosed about the shared world. However, what we perceive as beautiful or meaningful is based on the world itself, based on how we, ourselves, make sense of the world.

Because of our common sense, we know that Leonardo da Vinci's painting of the Mona Lisa has been deemed beautiful by other agents who engaged with it. There are two possibilities: Either the Mona Lisa will not appeal to my taste, and therefore I will not engage with it, or it will appeal to my taste, and I will enter an engagement with it. If I engage with the Mona Lisa, based on my common sense, I will know that this painting was perceived as beautiful by people before me. However, this does not mean that I am going to perceive the Mona Lisa as beautiful. After my engagement with the painting, in my judgment of it, which starts from common sense, I might judge the painting as neither beautiful nor meaningful. This also would not change the fact that the Mona Lisa was deemed beautiful by people before me. In other words, what *poiesis* creates needs to be seen, used, and consumed by *praxis* to be meaningful and to be deemed beautiful too. Even though the cultural products of *homo faber* are based on the appearances of *praxis* in the world, their endurance, their maintenance, and their meaning depend on their appropriation by *praxis*. In other words, the cultural end products of *homo faber* can only be meaningful when they are appropriated as meaningful by *praxis*.

Artifacts (as in cultural products) are first a stable or durable form of common sense; there can be various meanings embedded in them, and they can be designed to amplify one meaning over the other. But how the world becomes politically meaningful as a result of the hermeneutic circle that *praxis* opens up is dependent on the acting agent carrying out *praxis*. Since, after *poiesis* creates, no meaning can be a constant, what the relationship between the acting agent and the artifact is depends on the acting agent themself engaging with the artifact and the agent's common shared world in which the artifact is embedded.

For Arendt, then, the cultural products of *homo faber* not only stabilize the appearance of *praxis* retroactively but also influence the further appearance of *praxis* proactively. This shows that political action does not only depend on the political actor alone; the political actor is surrounded and therefore influenced by the cultural products of *homo faber*.

Conclusion

In this chapter, I have shown that *praxis* and *poiesis* are intertwined rather than opposed. Through an analysis of Arendt's notion of common sense, I demonstrated that Arendt positions common sense as both "world building" and "world disclosing" (Borren, 2013). This suggests a closer relationship between *poiesis* and *praxis* than Arendt first proposed. I showed that work, which corresponds to the human state of worldliness, creates the world, and I distinguished Arendt's construction of different areas that comprise the world. Following these distinctions, I argued that the world Arendt sees as encompassing the realms of the private and the public is indeed artificial, but it is also material. Even though Arendt does value *poiesis* and its importance for *praxis*, she is very critical toward one manifestation of the end result of the work of *homo faber*: technologies. Arendt finds world alienation to be the end result of the rapid technological advancements in her time. This leads her to situate technologies within the realm of the social and the accompanying mode of worldlessness. However, she does not consider all end results of *poiesis* as irrelevant to politics. I argued that if Arendt's conceptualization of the cultural end products of *homo faber*, like stories and artworks, is considered as a point of departure, rather than her limited reading of technologies, she can guide us toward an interaction-based politics of technology. Following this, I proved that the cultural products of *homo faber* not only stabilize the appearances of *praxis* but also inspire further appearances of *praxis*, because these products are an embodiment of the meaning that the world has been given at their time of creation.

In order to expand Arendt's insight into the political value of cultural artifacts for *praxis*-based political theory, a postphenomenological approach is needed. Postphenomenology can help extend Arendt's conceptualization of the cultural artifacts of *homo faber* to the material end results of the work of *homo faber*. In the next chapter, by complementing Arendt's interaction-based political theory with postphenomenology, I show that material end results of the work of *homo faber* are also capable of stabilizing the appearance of *praxis* retroactively and influencing its further appearance proactively, just as do *homo faber*'s cultural end products.

Notes

1 Borren explains subject–object dualism as entailing "the separation and opposition of, on the one hand, a subject, perceiver or self, and, on the other hand, an object or perceived, i.e. world and nature, including one's own body, other human beings, etc." (Borren, 2013, p. 233).

2 Arendt's italics.

3 As defined in Chapter 1, conditions include within themselves an openness, in the sense that even though we are conditioned beings, this does not mean that a certain condition is realized. In Borren's words, "a condition is not a causal mechanism or a one-way determination. By opening up the horizon of the future, the human conditions preserve freedom … The phenomenological background of the idea of conditionality ensures that conditions are both constants of human experience and existence and historically variable in their particular constellations and combination" (Borren, 2013, p. 237).

4 Italics added by Loidolt (Loidolt, 2018).

5 Marx's alienation theory is close to Arendt's reading of capitalism and its alienating impact, but Arendt's criticism of Marx remains intact. According to Arendt, Marx, like others, considered the things that we create as durable, tangible things that stabilize our lives and have an objective character as parts of our economy, as things to be bartered and exchanged. For Arendt, because of this perception, Marx remains confined to the modern age's extreme subjectivism because he attributes the same values to things as to owners of the capital (Arendt, 1998, p. 254). For more on Marx's alienation theory, see Marx (1972).

6 Pitkin also argues that "Arendt never advocates dismantling our economy, abolishing trade and money and reverting to household production for use" (Pitkin, 1998, p. 12).

7 The terms "mass society" and "society" in Arendt's terminology are two words defining one thing: a homogeneous group of people who are alienated from the world and only concerned with themselves. Thus, the terms "mass society" and "society" are also used interchangeably in this book.

8 Arendt's italics.

9 For an extensive reading of Arendt's response to the question of adequate housing, see Hill (1979) and Bernstein (1986).

10 While acknowledging the significance of the demarcation between the social and the political, I do not wholeheartedly exclude the social from the political.

Part Two

3

Technological Mediation of Common Sense

In the previous chapter, I showed that Arendt positions common sense as both "world building" and "world disclosing" (Borren, 2013). This, I argued, suggests that *poiesis* (world building) and *praxis* (world disclosing) are intertwined. I showed that despite adopting a rather pessimistic (almost technologically deterministic) attitude toward the material end results of *poiesis*, namely technologies, Arendt hints that the cultural end results of *poiesis* have influence on the meaning-generating capacity of *praxis*. I have shown that Arendt does not consider all end results of *poiesis* as irrelevant to politics. I demonstrated that for Arendt, the cultural products of *homo faber*, such as stories or artworks, are the embodiment of the common sense of their time of creation. Thus, the cultural products of *homo faber*, while stabilizing the appearance of *praxis* retroactively, influence the further appearance of *praxis* proactively.

In this chapter, I show that material end results of the work of *homo faber* are also capable of stabilizing the appearance of *praxis* retroactively and influencing the appearance of further *praxis* proactively, just as the cultural end products of *homo faber* are. In demonstrating this, I use this chapter to provide the first step in technopolitical mediation, namely technological mediation of common sense.

In the first section of the chapter, I show that it is necessary to make a move from the phenomenological tradition to which Arendt belongs to a postphenomenological approach. Postphenomenology shows that things (as in artifacts, objects, and technologies) have a hermeneutic dimension, and because of that quality, they mediate human experience and how humans have access to their world—in other words, how humans make sense of their world. Thus, following Verbeek (2005), I argue for a coshaping relationship between human action (*praxis*) and technologies. Following on this insight, I refer to Olya Kudina's (2019) scheme of the hermeneutic lemniscate to explore the mediation of common sense by technologies. I demonstrate that Kudina's hermeneutic lemniscate presents a new principle of human interpretation and sense-making. Therefore, by demonstrating that human interpretation and sense-making are technologically mediated, the model helps in explaining the role of the cultural and moral history of the person engaging with the technology in question.

In the second section of the chapter, following the (post)phenomenological turn to empirically informed microlevel studies, I conduct an empirical exploration. This allows me to investigate my proposed expansion of Arendt's political hermeneutics to materiality using a real-life example: Gezi Park, a public park in Istanbul. I investigate the active role played by the park in the instigation of the Gezi Park protests and the formation of a heterogeneous public during the protests. In this section, I describe in detail the methods followed to carry out this empirical exploration.

In the third section of the chapter, I combine the proposed expansion of Arendt's political hermeneutics to materiality with the results of the aforementioned empirical exploration. By doing so, I show that technologies mediate common sense, since they are the material form of the cultural memory of the society in which they are embedded. By providing extracts from the interviews that I conducted with the participants of the Gezi Park protests, I situate this public park as a space that has a place in the common sense of the people of Turkey. I show that Gezi Park is a political space for Turkish people because of its meaning within the common cultural context of the people. Following from this, I demonstrate that the people of Turkey resisted the impending demolition of Gezi Park because of its meaning and its place in their common cultural framework. Thus, I demonstrate that Gezi Park mediated common sense, as it is the material form of the cultural memory of Turkey.

From Phenomenology to Postphenomenology

In the previous chapter, I outlined the main contours of phenomenology while situating Arendt's political theory within hermeneutic phenomenology. In this section, I first show that while phenomenology bridges the gap between objects and subjects, postphenomenology further emphasizes this relationship by situating subjects and objects as coconstitutive (Verbeek, 2005, p. 112). Postphenomenology provides two approaches that enable understanding of this intertwined relationship between objects and subjects. These are hermeneutic or perception-oriented postphenomenology, which I examine in the second subsection, and existential or *praxis*-oriented postphenomenology, which I examine in the third subsection. This overview of the approaches within postphenomenology allows me to argue, in the final subsection, for a coshaping relationship between humans and technologies. I thus expand Arendt's insight into the political role of cultural artifacts to apply to material and physical artifacts as well.

Beginnings of Postphenomenology

Classical philosophy of technology, as mentioned in Chapter 2, is considered to be "backward-looking" (Verbeek, 2005, p. 8). Verbeek, instead, proposes a forward-looking approach that focuses on the concrete technological artifacts and objects themselves. This approach thus endorses empirical research into technologies while keeping in mind the existential and hermeneutic dimensions of analysis elaborated by classical philosophers of technology. In advocating this approach, Verbeek moves beyond analyzing the conditions of technology, arriving at an analysis of technology itself (Verbeek, 2005) to focus on the role technologies play in our worldly existence. By focusing on the things themselves, Verbeek moves beyond the phenomenological investigations of Jaspers and Heidegger through adapting Don Ihde's (1993) "postphenomenology" of technology. Postphenomenology is the incorporation of material hermeneutics to classical phenomenology. Material hermeneutics is the acknowledgment that material artifacts have a role in human interpretation of objective reality (Ihde, 2009). In other words, material hermeneutics would not treat the process of interpretation as a direct encounter between a person and the world but as a technically mediated one. In that sense, postphenomenology shows that there is no human existence without technology, and postphenomenology of technology presents humans and technologies as mutually interdependent in acquiring their characteristics. Verbeek argues that in the past decade, philosophy of technology has concerned itself with a central theme of mediation to examine the role of technological artifacts in society and how they influence human behavior (Verbeek, 2006). According to Verbeek, the framework of "mediation" analyzes artifacts "in the relationship between human beings and their environment" (Verbeek, 2006, p. 53). The mediation approach assumes that subjects and objects are not only intertwined but also coshape one another (Verbeek, 2005). This means that

> Human beings can only experience reality by relating to it, which does not involve any reality-in-itself but rather reality-for-them. As consciousness (perception, experience) can only exist as consciousness of something, reality is always reality for someone; in their engagement with reality, human beings always disclose it in a specific way. At the same time, humans themselves are constituted in this relation. The environment with which they are involved always codetermines in which ways they can be present to the world and each other. In the encounter between humans and world, each manifests itself in a particular way. In the mutual relation of humans and world there arises, therefore, a specific "objectivity" of world and a specific "subjectivity" of human beings.
>
> (Verbeek, 2005, p. 112)

This perspective allows one to acknowledge that the specific characteristics of both subjects and objects are to be disclosed in the relationship that they have with one another. In that sense, there is always mutual interdependency between subjects and objects:

> Humans and their world are always interrelated. Human beings cannot but be directed at the world around them; they are always experiencing it, and it is the only place where they can live their lives. Conversely, the world only gets a meaning for human beings in the relationships they have with it: it needs to be perceived and interpreted in order to be meaningful. Humans and their world, therefore, determine each other in the relations and interactions that exist between them. In their interrelation, both the subjectivity of humans and the objectivity of the world are shaped.
>
> (Verbeek, 2006, p. 54)

Don Ihde, in *Postphenomenology* (Ihde, 1993), translates the "intentionality" of consciousness to artifacts and argues that technologies "have a certain directionality, an inclination or trajectory that shapes the ways in which they are used" (Verbeek, 2005, p. 115). Verbeek explains this directionality as follows:

> Because of the intentional structure of human experience … humans are always directed toward reality. They cannot simply "think" but they always think *something*; they cannot simply "see," but they always see "*something*"; they cannot simply "feel" but always feel "*something.*"
>
> (Verbeek, 2008a, p. 388)

Human beings cannot just exist in themselves in the world; they experience the world, and when this experience begins, argues Verbeek, humans have no choice but to be directed to the things that constitute their world (Verbeek, 2008a).

Postphenomenology can be said to be the study of this relationship between humans and the things that constitute their world.

As mentioned before, the hermeneutic or existential scheme that phenomenology is concerned with is translated into a scheme in postphenomenology too. Accordingly, postphenomenology has two directions of analysis, corresponding to the hermeneutic and existential approaches of phenomenology. The first direction, "hermeneutic or perception-oriented postphenomenology" (Verbeek, 2006, p. 56), concerns itself with understanding the role of technologies in humans' access to the world and how these technologies influence human experience. The second, "existential or praxis-oriented approach" (Verbeek, 2006), concerns itself with analyzing the role of technology in how "human beings act in their world and realize their existence" (Verbeek, 2006, p. 56). In the following subsections, I provide an overview of both approaches within postphenomenology.

Hermeneutic or Perception-Oriented Postphenomenology

Hermeneutic or perception-oriented postphenomenology entails four perspectives that compose this approach: microperception, macroperception, forms of engagement with technologies, and multistability. According to Verbeek, Ihde, who focuses on the central phenomenological theme of "experience" that happens in the world,[1] and moves toward an investigation of the structure of this experience of the world (Verbeek, 2005, p. 122). Ihde argues that this investigation should be a "relativistic" (Ihde, 1990, p. 23) one, where the focus should be on the relations that emerge between humans and the world when the two engage with each other (Verbeek, 2005).

Ihde claims that this relational character of human–world interactions is the "ontological feature of all knowledge, all experience" (Ihde, 1990, p. 25). As mentioned before,[2] phenomenology recognizes that humans always experience something-as-something, and this makes the human who experiences and the thing that is experienced "inextricably bound up with each other" (Verbeek, 2005, p. 122). The process of interpretation allows one to discern and grasp the meaning of that something. Ihde argues that technologies are always in the way of this human interpretation of the world (Ihde, 1998).

a) Microperception and Macroperception

According to Ihde, there are two forms in which this process of human experience and interpretation takes shape. Microperception is a perception based on bodily senses (Ihde, 1993), and macroperception is a perception based on the culture in which one is embedded (Ihde, 1993). Ihde indicates,

> What is usually taken as sensory perception (what is immediate and focused bodily in actual seeing, hearing, etc.), I shall call microperception. But there is also what might be called a cultural, or hermeneutic, perception, which I shall call macroperception. Both belong equally to the lifeworld. And both dimensions of perception are closely linked and intertwined. There is no microperception (sensory-bodily) without its location within a field of macroperception and no macroperception without its microperceptual foci.
>
> (Ihde, 1990, p. 29)

According to Verbeek, microperception and macroperception can be distinguished, but they cannot be separated from one another. The reason for this is that "bodily perception can no more exist without being interpreted than an interpretation can exist without something to be interpreted" (Verbeek, 2005, p. 123). This means that "I cannot perceive a tree (microperception) if I do not have the basic, preliminary knowledge from my cultural background that the tall thing in front of me is called a tree (macroperception)"[3] (Verbeek, 2005, p. 123).

b) Forms of Engagement with Technologies

Ihde distinguishes four types of relations between humans and technologies when humans are making sense of the world. They can be summed up as follows:

> Embodiment Relation: Technologies that embody their users;
> Hermeneutic Relation: Technologies that provide representations of reality;
> Alterity Relation: Technologies that are a termini of human experience;
> Background Relation: Technologies that are not experienced directly but create a context for human perception.
>
> (Verbeek, 2008a, p. 389)

To illustrate these relations with examples, an embodiment relation would be wearing glasses: The user does not necessarily look at the glasses but looks at the world through the glasses. When used to perceive the world, the glasses disappear from the viewer's attention (Verbeek, 2005). In the case of hermeneutic relations, the technology is not transparent, as in the example of glasses (Verbeek, 2005). In hermeneutic relations, the technology provides a representation of the world, amplifying one aspect of it over others (Verbeek, 2005). This representation of the world through the artifact, however, requires interpretation; otherwise, it would not impart anything to us about the world (Verbeek, 2005). Hermeneutic relations thus involve interpretation, and, according to Verbeek, because the artifact in this relation must be read, Ihde rightly expands the hermeneutic relations humans have with texts to technological artifacts as well. In these relations, "the world is not perceived through the artifact but by means of it" (Verbeek, 2005, p. 125). The hermeneutic relation could be illustrated with the example of using a thermometer (Verbeek, 2005). The thermometer represents the temperature of the room, thus amplifying one aspect of it. When I read the thermometer, I am not necessarily involved with the thermometer but with what it reveals about the world, in this case the temperature (Verbeek, 2005). The third form of relations, namely the alterity relation, occurs when humans approach a technology in anthropomorphic ways—in other words, when they project human properties onto technologies, objects, or artifacts (Verbeek, 2005). Even though we know that a technology is never an "other," we still "care for" technologies (Verbeek, 2005, p. 127) or call them "smart" phones. The fourth form, the background relation, occurs when a technology is at work in the background of our experiencing of the world: for example, the central heating system or refrigerators. These technologies work, switching on and off in the background of our environment, and we scarcely recognize them while we are engaging with the world (Verbeek, 2005).

Once it is accepted that technologies are always in between humans and the world, it can also be claimed that humans' perception of the world is always transformed by technologies in one way or another. These technologies, however, are *multistable* (Ihde, 1990, p. 150), meaning that they are ambiguous and have multiple dimensions. This means that technologies "are never in themselves" (Verbeek, 2005, p. 117). What a specific technology is can be observed only in relation to the human who engages with it.

c) Multistability

Multistability affirms that technologies are only "stable" when they are being appropriated by a subject. However, the interpretation of the subject using the technology does not fix a meaning onto the technology; it only fixes a certain meaning during that certain use of the certain subject. The same object or technology can be used "in order to" do something else by a different subject. This would open different contexts of interpretations and meanings. Verbeek points out that the multistability of technologies does not exist "within a single use context, but always has a cultural index as well" (Verbeek, 2005, p. 136).

Verbeek continues:

> Technology is never purely determinative, for in principle other cultural relations with a given artifact are always possible. But neither is it purely instrumental, for when an artifact receives a particular definition within a cultural context—and thus becomes stable rather than multistable—it still contributes to shaping that context. Precisely because technologies are always interwoven with culture, they are always in a position to transform that culture—not "in themselves," but from the position that the cultural definition has given them.
>
> (Verbeek, 2005, p. 138)

Verbeek points out that cultural intentionality within the relation between humans and technology arises "when a cultural relation with an artifact is initiated" (Verbeek, 2005, p. 138). This occurs depending on what indirect form is given by the technology to the interpretations and experiences of human beings while also directly mediating their sensory perceptions (Verbeek, 2005).

In sum, hermeneutic or perception-oriented postphenomenology considers technologies as "means by which our perceptions and our wider experience are modified and transformed" (Ihde, 1998, p. 1). These relations all reveal that there is no human engagement with the world without the involvement of technology. However, one needs to be able to take into account "direct or indirect" (Verbeek, 2005, p. 128) ways of technological mediation of our perception of the world. For Verbeek, the direct way concerns the microperceptory level already mentioned, where technologies shape the way in which humans perceive reality and reveal possible ways in which reality can be interpreted (Verbeek, 2005). The indirect way concerns the macroperceptory level, in other words, the technological mediation in the frameworks of interpretation that coshape the relation between humans and their world (Verbeek, 2005). Verbeek, then, with his technological mediation theory, proposes a new insight within postphenomenology. As mentioned before, technological mediation theory asserts that humans and technologies coshape one another, a phenomenological recognition with which Verbeek's main purpose here aligns. Thus, Ihde's formulation that there are always technologies in between humans and the world is expanded by Verbeek with the phenomenological insight of the coshaping aspect of humans and the world. According to Verbeek (2005, 2011), Ihde fails to recognize the coshaping relationship between humans and technologies, which is a failure that would bring us back to the subject–object dichotomy that phenomenology aims to overcome. Thus, Verbeek argues that, even though it is tempting to perceive mediation as a form of transformation that occurs when a subject experiences an object with technology situated in the middle, this understanding can only be understood as the world: the object as a fixed object and the subject as a fixed subject (Verbeek, 2005).

However, phenomenology teaches that there is no fixed world and no fixed subject. Mediation, from a postphenomenological perspective, is not to be found only in the transformation of the world and the subject. Hermeneutics is in the meaning that is generated within the interaction between the object and the subject. When one tries to grasp meaning and understand the relation between subject and object, one needs to move beyond the subject and object themselves and grasp how they are mutually constituted in their interrelation (Verbeek, 2005). Verbeek argues that

> In any relation between subject and object, both are brought into existence in a specific way, and both subjectivity and objectivity acquire specific shape. When analysing the mediating role of artifacts, therefore, this mediation cannot be regarded as a mediation "between" subject and object. Mediation consists in a mutual constitution of subject and object.
>
> (Verbeek, 2005, p. 129)

By presenting the search for mediation of artifacts in the coconstitution of subjects and objects, Verbeek highlights that there is no pregiven subject or pregiven object that is in engagement with one another. Who and what the subject is, and what the object is, is constituted within the engagement—within the interaction between the two. As Verbeek argues, "what the world 'is' and what subjects 'are,' arises from the interplay between humans and reality" (Verbeek, 2008b, p. 13). Mediation, continues Verbeek,

> does not simply take place between a subject and an object, but rather coshapes subjectivity and objectivity. Formulations in terms of the "access to reality" offered by an artifact should be read as relating to the way in which an artifact makes possible the constitution of a world in the very process of perception. Humans and the world they experience are the products of technological mediation, and not just the poles between which the mediation plays itself out.
>
> (Verbeek, 2005, p. 130)

This quotation also suggests a move on Verbeek's side beyond the linear human–technology–world relationship Ihde suggests, toward a more circular structure. According to Verbeek, technologies can transform our perceptions (Verbeek, 2005).

Hermeneutic or perception-oriented postphenomenology, then, is the recognition that when humans interpret and experience the world, this always happens through the mediation of technologies. Humans, technologies, and the world are all active in the process of human interpretation and human experience. By pointing out that it is not only humans and the world coconstituting one another through technologies but also technologies and humans coconstituting one another, Verbeek adds a circular dimension to the linear human–technology–world relations as proposed by Ihde.

Existential or Praxis[4]-Oriented Postphenomenology

According to Verbeek, technologies not only mediate human perceptions and interpretations but also indirectly influence human actions (Verbeek, 2006, p. 57). Verbeek argues that "things are able to exert influence *as material things*, not only as carriers of *meaning*" (Verbeek, 2006, p. 58). According to Verbeek, human action is influenced by human intentions, the social structures in which one finds oneself, and the material environment in which one lives one's life (Verbeek, 2011). Following this recognition, Verbeek adapts Latour's concept of "script" while describing the technological mediation of action. Scripts describe the way in which an artifact is going to be used when a human engages with it (Verbeek, 2011, p. 10). A plastic cup, for example, has the script of "throw me away after use"; it is embedded in the design of the artifact that it shall be thrown away after use (Verbeek, 2006). Scripts here are defined as a framework that is inscribed into an object by its designer or architect to prescribe "how to act" (Verbeek, 2006, p. 58) when the object is in use.

While Verbeek recognizes that not all scripts are the end result of deliberate inscription, even those that are can end up influencing human behavior in an unintended way (Verbeek, 2006). Verbeek argues that in both mediations, namely mediation of perception and mediation of action, "*transformations* occur" (Verbeek, 2006, p. 59). In the terminology of mediation of action, these transformations are defined as invitation and inhibition. This means that, through design, technologies can invite certain human actions and inhibit others (Verbeek, 2011). Once the technology and human enter into a relationship, they become a "composite actor":

> By entering a relationship with another entity, the program of action of the original actor is translated into a new one. When somebody's action program is

> to "prepare meals quickly," for instance, and this program is added to that of a microwave oven, the action program of the resulting, "composite actor" might be "regularly eating instant meals individually."
>
> (Verbeek, 2006, p. 59)

Here, Verbeek points to an important difference between mediation of action and mediation of perception. Verbeek explains this with an example of the bulky key ring given to hotel guests, which invites the guests to return the key when they are leaving the hotel. Putting this example in postphenomenological terms, Verbeek argues:

> When hotel guests have the room key in their possession, the bulky attachment mediates the way in which they experience the key—namely, as something annoying—and therefore the way they deal with the key—namely, returning it to the reception on their departure. In the latter case, there is a translation of action, a displacement or transposition of the way in which humans are present in and deal with their world. Artifacts co-shape the use that is made of them, and thereby the relations that arise between humans and their world. In this mediation emerges a translation of action, which parallels the transformation of perception.
>
> (Verbeek, 2005, p. 171)

Verbeek here shows that mediation of action has a similar structure to the hermeneutic mediation of technologies: Just like transformation of perception through amplification and reduction, transformation of action occurs through invitation and inhibition. This means that technologies promote certain human actions and discourage certain other human actions (Verbeek, 2005, p. 171). Just like the hermeneutic mediation of technologies, mediation of action also mediates the "constitution of objectivity and subjectivity" (Verbeek, 2005, p. 171). By inviting and inhibiting certain human actions, technologies shape how one acts within the world and realizes one's existence. As Verbeek argues, "because mediated actions make humans encounter the world in a particular way, the mediating artifact helps to determine how both the world ('objectivity') and those who act in it ('subjectivity') are present" (Verbeek, 2005, p. 171).

Now that I have explained the two components of technological mediation as developed by Verbeek, it is possible to present a graphic representation of the theory.

Figure 1 clearly shows how Verbeek's technological mediation construct approaches human–technology relations. The straight line that is leaving the

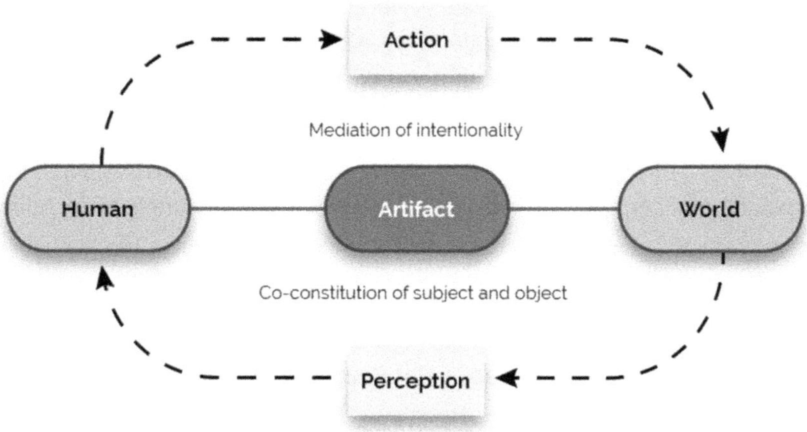

Figure 1 Technological Mediation Theory Scheme.

human and going toward the world is met by the technology or artifact on its route. This technology either amplifies or reduces a certain meaning within the world. This, in turn, influences how the human perceives and experiences the world. Technology also invites and inhibits certain human actions and practices. This, in turn, either limits or enables the range of activities that humans can carry out within the world. Through this scheme, Verbeek portrays that (a) technologies mediate human perceptions and (b) technologies mediate human actions.

Olya Kudina (2019) is skeptical of the extent to which Ihde's material hermeneutics, which has been incorporated into this technological mediation scheme, is explanatory of the complex relationship between humans, the world, and technologies in the process of interpretation. According to Kudina, the existing scheme needs to be expanded, since technological mediation as is does not "explain in an encompassing manner the interpretative processes that underpin human understanding in an encounter with a (new) technology" (Kudina, 2019, p. 101). According to Kudina, to enable a grasp of the "mediating role of technologies in the process of interpretation as well the productive nature of the historical horizons" (Kudina, 2019, p. 102), the material hermeneutics scheme of technological mediation should be combined with the scheme of the hermeneutic circle. In the subsection below, I explore in depth what Kudina's addition of the hermeneutic lemniscate to the technological mediation theory scheme entails for the creation of a new account of technological mediation of politics.

The Hermeneutic Lemniscate

In this subsection, I refer to Olya Kudina's (2019) scheme of the hermeneutic lemniscate to explore the mediation of common sense by technologies. I show that Kudina's (2019) hermeneutic lemniscate, expanding on the material hermeneutics of Don Ihde (1990), showcases a new principle of human interpretation and sense-making, namely that human interpretation and sense-making are technologically mediated.

According to Kudina, Ihde's material hermeneutics, while acknowledging the mediating role of technologies in the process of interpretation, does not explain how the conveyed reality returns to the existing experiential and sociocultural perceptions of people (Kudina, 2019). Kudina argues that the coconstitution aspect that Verbeek (2005) brings to postphenomenology still fails to "account for how the prior perceptions and sociocultural awareness of people are co-shaped with technologies" (Kudina, 2019, p. 101).

To fill this gap, Kudina adds the hermeneutic circle to Ihde's material hermeneutics in an attempt to provide a comprehensive account of the interpretation underlying the way people appropriate technology (Kudina, 2019). To grasp the mediating role of technologies in the interpretation process, as well as the productive character of the historical horizons (Kudina, 2019, p. 102), according to Kudina, the material-hermeneutic scheme of technological mediation should be based on the scheme of the hermeneutic circle. Kudina explains:

> Considering the mediating role of technologies in the process of interpretation as well the productive nature of the historical horizons that each of the components in the process of interpretation inalienably possesses, a hermeneutic situation will resemble a combination of two hermeneutic circles, interrelated and always in flux.
>
> (Kudina, 2019, p. 102)

Therefore, Kudina develops the hermeneutic lemniscate in order to be able to explain value change and explore how humans appropriate new technologies even before those technologies enter the world of human life. Kudina here points out the ephemeral and dynamic structure of meaning making once one encounters a technology. The hermeneutic lemniscate is formulated and explained by Kudina as "a figure-eight shaped curve (∞) consisting of three linking interrelated components: human, technology, world" (Kudina, 2019, p. 102). The lemniscate model figuratively follows the dynamic and interrelated entanglement of humans (H), technologies (T), and the surrounding world (W) in the process of technological appropriation (Kudina, 2019). The structure of the lemniscate itself is visually represented by a model with arrows navigating back and forth between a person, a technology, and the world. This model calls for the inclusion of technological practices and of the ways in which people foresee the use of a particular technology in a particular environment (Kudina, 2019).

The hermeneutic lemniscate explores how humans appropriate new technologies, even before those technologies enter the world of human life. Therefore, the model helps in explaining the role of the cultural and moral history of the person engaging with the technology in question. Kudina provides the example of a person engaging with an e-reader. Once this person (the human in Figure 2) encounters this technology (the e-reader), she tries to understand what she is confronted with, drawing on all the preexisting structures of interpretation that are present, which are based on a greater historicity (Kudina, 2019). The symbol in the middle (technology as the e-reader) is a technologically mediating actor in the process of interpretation, which necessarily frames reality in a certain way, suggesting certain perceptions and actions while obscuring others (Kudina, 2019). The symbol on the right represents the world event that a person is confronted with, in this case through a story in an e-book. The story projects certain meanings and perceptions onto the reader, which the reader has to interpret. She does this in a dynamic process of panning between her existing interpretive structures and the projected meaning of the story, with the e-book always acting as a mediating link between the two (Kudina, 2019).

Using this example, Kudina argues that when the process of interpretation passes through all three parties—the reader, the device, and the world event described in a story—temporary stabilization of meaning occurs (Kudina, 2019). There is a constant back and forth between the initial request of the reader, the

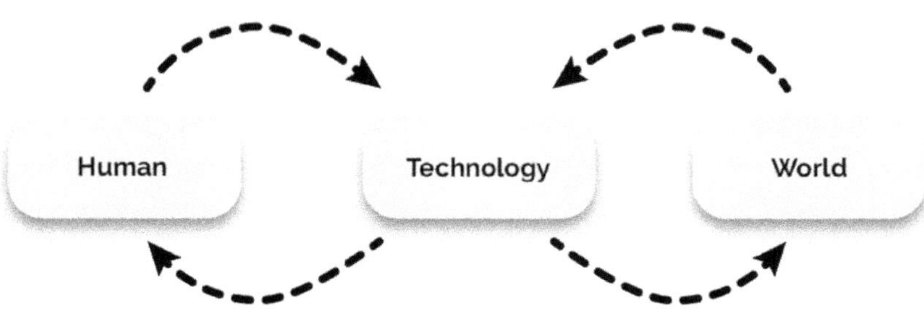

Figure 2 The Hermeneutic Lemniscate.

technological mediation of the e-book, and the story that depicts certain events in the world (Kudina, 2019). This constant back and forth, according to Kudina, represents the merging of the horizons represented by each of these parties and creates a sense-making activity that is led by temporary understanding (Kudina, 2019). With this prior understanding, both the projective and the practical appropriations of the e-book stabilize until they can no longer do justice to the changes in the person's environment, the technological equipment, or the person herself (Kudina, 2019). When this happens, the tentative projective meaning or defined practical utility of a technology will reflect these changes, with the mode of appropriation changing accordingly (Kudina, 2019). As such, the process of appropriation never ends and reflects the fluid, interconnected model of the lemniscate elements (Kudina, 2019).

With the hermeneutic lemniscate, Kudina shows that any hermeneutic inquiry should always simultaneously question what the technology and the world situation are projecting and how these reflect the existing structures of understanding of the respective person (Kudina, 2019). The three parties are never separate but always involved in the interpretation process together. What Kudina highlights is a specific hermeneutic situation that needs to be examined from all sides, regardless of the initial entry point into that situation (Kudina, 2019). Kudina argues that the lemniscate structure of the interpretation ties in with the technological mediation approach, an approach that suggests that perception and action are closely linked (Kudina, 2019). In the lemniscate, too, the perceptions that the reader receives when encountering a technological form the background for further action, which stimulates the reader either to follow the suggested meaning or to refute it (Kudina, 2019). When absorbing an item of knowledge conveyed by the book, the reader always correlates the incoming information with the earlier ideas and experiences that underpin her preliminary structures of understanding (Kudina, 2019). In parallel, the reader's physical perception of the world also plays a role. For example, it affects how the statements in the book reflect the current situation in the surrounding world and the views, not only of herself but also of her family and friends, that require an action by the reader (Kudina, 2019). This can inspire a new meaning of the book based on the effective story of the reader, the mediation ability of the e-book, and the proposed narrative of the book (Kudina, 2019).

Kudina focuses on the moral change and the change in values in society once the process of appropriation (meaning making) begins (Kudina, 2019). She suggests viewing the moral landscape of people as part of the effective history that enables the appropriation of technology (Kudina, 2019). The lemniscate model makes it possible to distinguish an inalienable hermeneutic dimension of technological mediation from morality (Kudina, 2019). In the lemniscate, the productive history of humanity projects its initial prejudices about a new technology and confronts the normativity of technology, a conflict which is staged in certain practices through its design and actor intentions (Kudina, 2019). The interpretation process denotes an encounter and a revision of the originally projected values in the merging of several horizons (Kudina, 2019). The meaning of values stabilizes temporarily, whereby the originally projected values are either confirmed, revised, or supplemented with new normative levels (Kudina, 2019). The values at stake remain dynamic in the face of the continuous process of interpretation. This process is guaranteed by the constantly expanding preexisting structures of human interpretation and contextual factors, and it is conveyed at the same time by the existing technology (Kudina, 2019). The technically mediated hermeneutic lemniscate reflects all these insights in their flowing and continuous movement between the human, technological, and secular parties in the process of interpretation (Kudina, 2019).

With the model of the lemniscate, Kudina (2019) shows that it is both the world and the technology that play a role in how we make sense of things. In this scheme, the world, as much as the technology, is an active component in sense-making. I argue that the model developed by Kudina allows the extension of Arendt's understanding of the implications of the cultural products of *homo faber*—as the embodiment of the common sense of their time of creation—to material artifacts as well. This would also mean extending the influence of the cultural products of *poiesis* on the further appearances of *praxis* to material artifacts as well. Surprisingly, Arendt herself on one occasion acknowledged that our common sense is also informed by physical, material products like public parks or buildings; however, she left this insight underdeveloped. Arendt writes:

> If … we understand the political in the sense of the polis, its end or raison d'être would be to establish and keep in existence a space where freedom as virtuosity can appear. This is the realm where freedom is a worldly reality, tangible in words which can be heard, in deeds which can be seen, and in events which are talked about, remembered and turned into stories before they are finally incorporated into the great storybook of history. Whatever occurs in this space of appearances is political by definition, even when it is not a direct product of action.
>
> (Arendt, 2006a, p. 153)

Following this, the model of the hermeneutic lemniscate, once added to the insight provided by Arendt, allows an acknowledgment that there is constant interplay between *praxis* and *poiesis*, and it is this interplay that adds to humans' common sense, which is the basis for all understanding. Following the hermeneutic lemniscate, such an argument for an interplay between *praxis* and *poiesis* in relation to meaning making does not subordinate one activity to the other. Because both *praxis* and *poiesis* are within the world, and because they are both *for* our shared common world—and by extension our common sense—the way in which we make sense of the world is an end product of the interplay between these two activities. We know that the meaning technology mediates is not static, and for technology to continue mediating meaning, it has to be appropriated by *praxis*. In other words, technology has to be perceived, used, or consumed in relation to the meaning that was mediated through the technology. Technological mediation of common sense, then, acknowledges the following: (a) that how we make sense of the world is the accumulation of the records of a constant engagement between *praxis* and *poiesis*, between authentic appearances of acting agents and the material, physical, or cultural world; (b) that this engagement should be construed as "mediation," where *zoon politikon* and *homo faber* coshape one another through the activities they carry out within the world; and (c) that in this coshaping engagement between *praxis* and *poiesis*, not only the cultural products but also the material or physical products of *homo faber* are at play in how we make sense of the world.

Empirical Explorations

Now that I have demonstrated the concept and the structure that allow for an acknowledgment of the implications of the material products of *poiesis* on *praxis*, it becomes essential to study this technological mediation. Following the (post)phenomenological turn to empirically informed microlevel studies, and to elaborate this expansion of Arendt's political hermeneutics to materiality, an empirical exploration is necessary.

In this book, the version of empirical philosophy I conduct is relatively "lightweight" in that it does not use empirical data as a groundwork for conceptual elaboration. Rather, it uses elements of ethnography through the conversations with the protest participants, which are also supported by the existing literature, for illustrative purposes and to set the stage for the analysis. The modest version of empirical philosophy I offer here thus does not purport to provide normative guidance. Following Annemarie Mol (2021), I embrace the idea that the stories told by the interviewees do not lead to theoretical conclusions but rather challenge our understanding of community, politics, and interaction. According to Mol (2021), empirical philosophy interferes with established reflexes about theory and in that respect has not only brought philosophy "down to earth but also altered the empirical" (Mol, 2021, p. 8). In this exploration, I investigated the active role played by Gezi Park[5] (a public park in Istanbul) in the instigation of the Gezi Park protests and the formation of a heterogeneous public during the protests.

The empirical investigation in this book takes the form of a case study exploration of the role that Gezi Park played during the Gezi Park protests. These protests started on May 30, 2013, when Turkey's largest city, Istanbul, witnessed public unrest in its central area surrounding Gezi Park. Social media platforms—especially Facebook and Twitter—played an important role in the protests, providing a space for disseminating information (nationally and internationally), as well as facilitating communication among protesters (Chrona & Bee, 2017). Gezi Park and the 2013 protests became synonymous; hence, to understand how the dynamics of the protests proceeded, it is also important to understand what the park itself contributed to their development. In this way, the Gezi Park protests present a curious intersection between political theory and the philosophy of technology that includes not just the space for digital contestation but also the spatial and cultural arrangements that coshaped the protests.

There are various interpretations as to why the Gezi Park protests happened and what was their goal: the phenomenon has been seen as a protest against the neoliberal policies of the government (Gürcan & Peker, 2015), or as citizens exercising their right to claim the city (Kuymulu, 2013), while some analyses have focused on class structure and the emergence of new classes (Keyder, 2014), others have considered public space and the importance of such a space for self-expression and democracy claims (Göle, 2013a, 2013b, 2013c), and still others have viewed the protests as an attempted coup against the government and the rule of the Justice and Development Party (AKP) (Ete, 2013) or as a heterogeneous movement that aims to change existing structures (Özen, 2015). These analyses also highlight the plural, interactional aspect of political experience in Gezi Park during the protests. While highlighting the diversity of social groups during the protests, these analyses all argue that the political value of the protests was in the struggle for power between the protesters and "the state," namely the ruling party, the AKP. It is true that these studies all acknowledge the protesters were a heterogeneous group of people, but they also see this plurality of perspectives within the park as a weakness of the protests (Özen, 2015). While I do not reject the validity and importance of any of these analyses, there is something that is missing in all these studies (and is still missing today). This is a thorough investigation into the active role of the material, physical public park itself in instigating the protests and generating this heterogeneous public.

In this investigation, I draw heavily on the existing literature analyzing the events of the Gezi Park protests (Baban & Güzel, 2015; Chrona & Bee, 2017; Eken, 2014; Haciyakupoglu & Zhang, 2015; Kuymulu, 2013; Tufekci, 2013; Tufekci, 2017; Özbank, 2013). Occasionally, I also draw on the semistructured interviews I conducted with the citizens of Istanbul and the participants of the Gezi Park protests to illustrate and qualify the conceptual analysis. Both aspects of the analysis here provide insights for further exploring technopolitical mediation.

Gezi Park and Technological Mediation of Common Sense

In this section, I combine the results of my empirical exploration with the proposed extension of Arendtian political hermeneutics to materiality. This allows me to argue that technologies mediate common sense since they are the material form of the cultural memory of the society in which they are embedded. I portray this argument in relation to the case study in three steps. First, I show that Gezi Park is a political space for Turkish people because of its meaning within the common cultural context of the people. Following this, in the second subsection, I show that the people of Turkey resisted the impending demolition of Gezi Park because of its meaning and its place in their common cultural framework. Thus, I demonstrate that Gezi Park mediated common sense, as it is the material form of the cultural memory of Turkey. In the third subsection, I show that the use of digital spaces was also influential in the technological mediation of common sense.

Gezi Park: History, Culture, and Design

Gezi Park is one of the most important public parks in Istanbul. There are several cultural and historical reasons for the park's importance: First, it is directly adjacent to Taksim Square, which is a concrete representation of the secular transformation of the country (Yildirim & Erdem, 2007). Second, the area that includes Taksim Square and Gezi Park has a special place in the memory of leftist groups in Turkey, since Taksim Square hosted the events of the Labor Day massacre of 1977. Third, Gezi Park was designed to implement Western-style patterns of socializing in Turkish society. The location of the park also makes it a significant space for public gatherings. According to one of the respondents,

> Taksim and Gezi Park is a place where one way or another your paths cross with these public areas. If someone comes to Istanbul for the first time, they go to Taksim, where shall I go? I should go to Taksim ... for example, a lot of friends of mine from university who came from Diyarbakir, Mardin, or Sirnak [cities in the east of Turkey][6] first went to Taksim the moment they came to Istanbul. Because Taksim is a very different place, it is the heart of Istanbul and everyone absolutely passed through that park too.[7]

Taksim Square is an inclusive space where daily activities, as well as leisure activities, are carried out by a vast diversity of people. According to Baykan and Hatuka (2010), Taksim Square is

> a place bustling with the movement of traffic and people, locals and tourists. Its vastness and facilities allow for different practices—sitting [near the monument] or in a café, shopping, eating and commuting. It is a place which maintains a reciprocal *daily* relationship between its space and the crowd's trajectories.
>
> (Baykan & Hatuka, 2010, p. 59)

In the space that is occupied by Gezi Park today, there used to be artillery barracks and its courtyard, built between 1803 and 1806, during the Ottoman period (Baykan & Hatuka, 2010). Baykan and Hatuka state that

> During the First World War, this installation was abandoned, and after 1921, the courtyard became a stadium, and the interior areas of the barracks were used as casinos and workshops. In 1940, the building was demolished to create an open space for a park. Apart from the barracks and the water reservoir, there were also cemeteries around the open space."
>
> (Baykan & Hatuka, 2010, pp. 53–4)

In time, the Beyoglu district, where this open space was situated, expanded, and new, upscale residences were built around the Taksim area, close to the lively Western lifestyle that people enjoyed in Beyoglu (Baykan & Hatuka, 2010). According to Baykan and Hatuka (2010), Taksim Square is a monument that signifies Turkey's "changing socio-political modes of operation" (Baykan & Hatuka, 2010, p. 51). Even though the transformation of Taksim Square can be traced back to the nineteenth century, Baykan and Hatuka argue that the space evolved as a monument during the transformations of the twentieth century and became a "concrete representation of the Turkish secular state" (Baykan & Hatuka, 2010, p. 51).

The physical transformation of the square and the construction of Gezi Park date back to the early twentieth century. In 1936, the municipality of Istanbul offered Henry Prost the task of turning the area that was left after the demolition of the artillery barracks into a public park. In 1939, Prost's master plan for Istanbul was endorsed by the Ministry of Public Works[8] (Gül, 2009). Even though it proved impossible to implement Prost's entire plan due to financial hardships caused by the Second World War (Gül, 2009), the transformation Prost wanted to implement in Taksim Square and its surrounding area was realized (Gül, 2009, p. 116). The plan was supposed to be in line with the agenda of the new secular republic, expressing the "discourses on republicanism, civic rights, progress, secularism and civilisation" (Baykan & Hatuka, 2010, p. 55). According to Gül, the reorganization of Taksim and its surrounding area was the centerpiece of Prost's master plan (Gül, 2009). Gezi Park was initially named "The Inonu Gezisi" (Inonu Promenade) after the second president of Turkey, Ismet Inonu (Gül, 2009, p. 116). The Promenade was established on a site of 26,000 square meters, "specifically planned as a walking route with grassed areas, flowerbeds, magnolias, trees, seating benches and kiosks for concerts of the Municipal Philharmonic Orchestra" (Gül, 2009, p. 116).

In time, Gezi Park became a central spot for spending leisure hours, in the middle of a busy area where one could relax and enjoy some free time. Akpınar and Şoher argue that "the geometrical design of the park, sequentially rooted up trees, carefully placed flower bulbs, the pool and the statues are pointing to more than an architectural choice"[9] (Akpınar & Şoher, 2013, p. 5). What this means is that the park was not created solely as a space for people to spend leisure time; the architect and the municipality of Istanbul had other things in mind. When journalists asked Prost why he was designing parks for Istanbul, he said that the reason he designed the parks was to take the women and children out of their cages (Akpınar & Şoher, 2013). Considering the time when this project was given to Prost, it should be noted that ten years before, in 1926, Turkey adopted the Swiss Civil Code and gave women "equality in civil rights" (Üngör, 1968). Two years prior to Prost's appointment, in 1934, Turkish women were given fully fledged citizenship, including the right to vote and be elected (Üngör, 1968). In Prost's master plan, the parks and promenades were designed for men and women who are participating equally in public life to enjoy social and recreational activities together, which would not be possible in the traditional *mahalle* (Gül, 2009). A *mahalle*

> is a space which extends the interior space of the family to the residential street; it is a space of belonging and collectivity. The most important practice for creating and sustaining the familiar spaces of *mahalle* life is neighboring (komsuluk), which makes home spaces open to neighbors. The cultural practice of neighboring is gendered, relying in part on traditional gender roles for women as wives and mothers which place them at home during the day.
>
> (Mills, 2007, p. 336)

The secular republic of Turkey, then, aimed at moving women beyond their traditional gender roles in *mahalles*, providing them with spaces to move about while endowing them with the freedom to do so through legal reformations. Prost's designs for public spaces, and his suggestions in the master plan for Istanbul, support radical social and cultural transformation for a mixed society[10] (Akpınar & Şoher, 2013, p. 5). The design of the park, including the sequential planting of trees, allows people to see others while walking along the path. At the end of the route, there is a circular area with park benches. This design was created to allow people, who are free to move about in the park, to encounter one another and possibly interact if they wish to do so.

Recreational areas in Turkey, which were created during the twentieth century, are not only a matter of adornment but an original feature of Turkish modernization (Akpınar & Şoher, 2013). Creating spaces as such points to a radical change in urban life, since this caused a new social life by changing the structural environment of the city (Akpınar & Şoher, 2013). According to Akpınar and Şoher, this plan illustrates that planning of a public space not only covers spatial elements but also allows men and women to freely enjoy mixed public life (Akpınar & Şoher, 2013). Thus, the plan becomes the legitimizing platform that allows theoretical reforms to be seen, both on a national scale that covers the conservative religious segment of society and on an international scale, especially by Western European countries (Akpınar & Şoher, 2013). This recreational and cultural space became the showcase of the new Turkish Republic (Akpınar & Şoher, 2013).

Taksim Square was also appropriated by different dimensions and groups of people over time. According to Gul et al. (2014), Taksim Square has hosted various labor demonstrations and Labor Day celebrations, as well as New Year celebrations and football celebrations. In short, it is the gathering spot for different interest groups to raise their voices, for whatever reason they may have (Gül et al., 2014). The square, and its surrounding area, has a powerful image and status in the minds of the Turkish people. As one respondent said,

> Taksim means freedom, I mean, if a girl from a conservative family goes to Taksim, it wouldn't be very welcomed, what does it mean that you are going to Taksim? I mean, because Taksim means freedom, it means a young person can walk around freely, sit in a park and drink a beer freely … Actually we shouldn't expect too much from a country where being able to sit in a park and drink a beer means freedom … but you don't think about this in Taksim, it is more like your own space of freedom, it feels like it is not tainted.

This response shows exactly why the Gezi Park protests happened *in* Gezi Park: because of how that space was designed and what had previously happened in that space; because of what that space means for people, what that space enabled to appear, how people's stories were told, and what these stories represent.

Whitehead and Bozoğlu (2016, p. 1) present the Taksim-Gezi area as a "site of ground memories," which are appropriated differently by different groups of people. Whitehead and Bozoğlu argue that Taksim Square is an "identity place," explaining that such a "place is formed into embodied historiography and material for identity construction" (Whitehead & Bozoğlu, 2016, p. 4). Such a place, according to Whitehead and Bozoğlu, "is remade as a source through which people may construct their own identities in relation to a politically-framed, geographically-located past, even where the 'past' in question extends back well before people's own lifetimes" (Whitehead & Bozoğlu, 2016, p. 4). What this means is that, when people associate a place with historical events that they define as significant for their identities, that place becomes an identity place, in Whitehead and Bozoğlu's terms (Whitehead & Bozoğlu, 2016, p. 4). An identity place is also called a "ground of memory," because these places housed various political activities, from state actions to political protests (Whitehead & Bozoğlu, 2016, p. 5). However, the memory housed by this ground is not one collective memory, but instead plural ground memories; these memories are vivacious, in the sense that a place can host different and even conflicting memories, which can lead to different place identities that are associated with those memories (Whitehead & Bozoğlu, 2016). According to Whitehead and Bozoğlu,

> Ground memories are evident physical traces, presentations, and constructions of place histories that enable, influence, or inform human behaviours, beliefs, and senses of belonging. The sense of "ground" here is both physical and metaphorical. It is the earth and what is buried in it, what is or was known to be built on it, was erased from it or took place on it.
>
> (Whitehead & Bozoğlu, 2016, p. 5)

Referring to Till's[11] work (2005), Whitehead and Bozoğlu align themselves with the conceptualization of places as having *relational agency*, since they are the scenes of experiences that enable and mediate social spaces and temporalities (Whitehead & Bozoğlu, 2016). This relational agency, for Whitehead and Bozoğlu, stems from the fact that these places carry within themselves the traces of interactions that happened in them (Whitehead & Bozoğlu, 2016).

I argue that Whitehead and Bozoğlu's understanding of Gezi Park as a ground of memory allows the articulation of Gezi Park as a space of appearance that is "political by definition … even if there is no action performed at the moment" (Arendt, 2006a, p. 153). As I mentioned before, Arendt, only once, acknowledged that some material, physical spaces carry the traces of the actions that happened in them before, in a similar way to cultural artifacts being the end result of *poiesis* (Arendt, 2006a). Turkish people know what Gezi Park means because of their shared common world, which relies on their common sense. The common sense of the time of its construction, which could be translated as becoming free citizens rather than the subjects of the Ottoman Sultan, was embedded in the park. In that sense, Gezi Park presents a unique case of how the common sense of the community can have a material representation. People are drawn to these physical public spaces because those spaces have meaning for them. People remember what happened to them, and those events mean something for the people as individuals and as a community. However, these memories are not constant; not everyone remembers the same thing related to the physical space, and not everyone engages in the same way, but everyone has some sort of engagement.

If it is possible to extend Arendt's articulation of cultural artifacts stabilizing the appearance of *praxis* retroactively to material artifacts, Gezi Park can be construed as such an artifact, based on its history and its space in the cultural framework of the community in which it is embedded. If this is so, I argue that the fact that cultural artifacts also influence further appearances of *praxis* prospectively can be extended to material artifacts, exemplified by Gezi Park, as well. The park itself is a space for freedom in the common sense of the Turkish people, and therefore, it inspires the political activities of the people who interact with it. In the following subsection, I explore the Gezi Park protests that took place in response to the proposed demolition of the park. In this exploration, I present the influence of Gezi Park itself, the material public park, on how the participants experienced the world before, during, and after the protests.

The Gezi Park Protests

In May 2013, civil unrest broke out in Turkey, initially to contest the urban development plan for Taksim Gezi Park in Istanbul. Before sunrise on May 31, police forces entered the park and violently dispersed the protesters who occupied the park in order to stop its demolition as part of a municipal urban renewal project (Gürcan & Peker, 2015). Before May 31 had ended, hundreds of thousands of people across the country took to the streets in a spontaneous and collective response to what they perceived as the government's increasing authoritarianism and conservatism (Gürcan & Peker, 2015). With this, a simple urban park was transformed into "a symbol of resistance against the increasingly authoritarian and socially interventionist rule" of the government (Gürcan & Peker, 2015, p. 7). During the protests, violent scenes and clashes between protesters and police officers erupted around Gezi Park and Taksim Square (Gürcan & Peker, 2015). By June 15, due to the excessive use of force by the police, the protesters had left the park (Gürcan & Peker, 2015).

The reason for the destruction of Gezi Park was the "Taksim Pedestrianization Project," which the ruling AKP government and the municipality of Istanbul wanted to implement. The project aimed at a total renewal of the Taksim and Gezi Park area. To provide a short background, since the general election of 2002, the AKP had been ruling Turkey as a single-party government with a specific focus on Islamic-based policies, such as ending the ban on headscarves[12] and supporting Imam Hatip (clergy) schools (Kaya, 2015, p. 50). It was only the third Islamist party in the modern Turkish Republic to become a part of the government (Kaya, 2015). After 2002, the AKP gained an absolute majority in parliamentary seats in the general elections of 2007 and 2011 and in the local elections of 2004, 2009, and 2014 (Kaya, 2015). Erdoğan was very successful in relating to people in "subaltern, conservative, and religious circles," who perceived him as one of them (Kaya, 2015, p. 50). Turkey has had a problematic relationship with Islam since the establishment of the modern Kemalist Republic. The abolition of the Caliphate of Islam and of the monarchy was implemented by an act of parliament, but this legal abolition did not mean that their demise was accepted by the people of the country (Ahmad, 1993). Modern Turkish history is full of moments where Islamic parties were shut down and governments abolished[13] (Kaya, 2015). The rule of the AKP became the antithesis to the secular Kemalist ideology in Turkey. Referring to their concept of rule as "conservative democracy," the AKP criticized all previous governments that had adopted Kemalist ideology in Turkey as "despotic, oppressive, imposing, homogenizing, proclamation from above, and social engineering" (Kaya, 2015, p. 54). Conservative democracy was proposed as being a counter to "political projects of social engineering" (Kaya, 2015, p. 54), allowing citizens religious freedom and thus aiming to represent excluded Islamic values (Kaya, 2015, p. 54). According to Kaya, the AKP aims at Islamizing Turkey by revitalizing "religiosity in the age of globalization" and by "highlighting the power of the individual, family, charity, faith-based organisations, community, and Islam" (Kaya, 2015, p. 63). Religious revivalism is also implemented by the AKP through attempts to criminalize adultery, lift the headscarf ban, reinforce familial values, and revitalize conservative values (Kaya, 2015). Additionally, it takes the form of Islamizing public spaces by means such as building mosques, turning churches into mosques, and attempting and succeeding in separating female and male students in certain schools (Kaya, 2015).

The Taksim Pedestrianization Project, which included the revival of the Ottoman artillery barracks in the space that was occupied by Gezi Park, was just one among many Islamization projects the AKP government had been carrying out. The project received much criticism from "architects, professional bodies, nongovernmental organisations and community groups" (Gül et al., 2014, p. 68), and most of this criticism was related to the lack of communication between these stakeholders and the municipality of Istanbul. According to Gül et al., one of the greatest objections to the project rested on the fact that not only would Gezi Park, a public space open to all for recreation, be demolished and replaced by a building, but also that this building would be a private enterprise—a shopping mall (Gül et al., 2014, p. 69). Some of the respondents perceived this pedestrianization project as an attack on the leftist and secular significance of Taksim. One respondent said,

> In that Taksim, in 1977, 34 people got killed … now, look, cities have squares that are the soul of them, as a political scientist I wouldn't say or write this … but I do think we can make such a comparison or use it as analogically Ms. Bas, Tahrir square in Egypt is such a square. For example Ankara has Kizilay, Izmir has Gundogdu … the thing is, what Tayyip Erdoğan did was an intervention in the collective past of the left.

The event the respondent refers to as a collective memory of the left happened on May 1, 1977, when thousands of people gathered to celebrate Labor Day in Taksim Square, and thirty-four were killed during the violent demonstrations that followed.[14] What was surprising and significant about the Gezi Park protests was the plural, interactional aspect of political experience. For example, Gürcan and Peker argue:

> what is to note as a unique characteristic of the GPPs[15] is the unlikely alliance and solidarity established among a diversity of groups. A week prior to the protests, the idea of bringing together such multitude of ideologies would be dismissed by any specialist in Turkey. The protest cycle took everybody by surprise. Groups with strong Kurdish and Turkish nationalist tendencies, secularists, and devoted Islamists, Feminist LGBT communities and groups with patriarchal mentalities, fans of antagonistic soccer clubs, various class factions, and so on unexpectedly joined together against the authoritarianism of AKP, and its leader in particular. We hold that the political significance and strength of GPPs stem largely from those unlikely alliances. These alliances … opened up the possibility of communication, empathy and solidarity between groups that were at best indifferent to each other.
>
> (Gürcan & Peker, 2015, p. 96)

The political environment of the country before the Gezi Park protests can be described as highly polarized. According to a survey and analysis conducted by Lindqvist and Östling (2010), Turkey ranked second highest in terms of political polarization among seventy-four democratic countries worldwide. It was clear from the respondents' answers that these people attended the Gezi Park protests not because they had a certain idea or ideology in their minds. They wanted to support the protesters within the park, and they wanted to be able to stop the demolition of the park, a decision that had been taken without them being consulted. They wanted to say that they were there, they existed, and their opinions mattered. As one respondent said,

> the park influenced my ideas about politics because you feel a very strong power, I mean I can do it, I can pour down the streets, that makes you realize that I have power, I don't belong to any political party, I haven't participated in anything before and there are a lot people like this, a very apolitical group of people but they are in the front lines, that allows you to be able to fight with your life on the line, it really makes you feel confident in yourself … in the sense that I have confidence in me and the people around me, I mean there is nothing secure about the situation but that was something I have never felt before.

What this respondent mentions is one of the surprising aspects of the Gezi Park protests: the gathering of a multiplicity of agents from various ideological backgrounds. Below, I provide the results of a survey carried out by Konda (2014), a research company in Turkey. I refer to Konda's research because it was conducted while the protests were still going on, whereas I was only able to interview the participants of the protests three years after the event had occurred. To avoid uncritically accepting accounts of the past that may have been romanticized by the respondents, I refer to Konda's research as a checkpoint to compare and contrast respondents' claims with the Konda survey results. According to Konda's research, only 21.1 percent of the protesters were members of a political party, association, or some similar kind of nonprofit organization (Konda, 2014). However, 31 percent would have voted for the Republican People's Party if there had been elections at that time (Konda, 2014). A further 29 percent were undecided, 18 percent would not have been able to choose a party, and 8 percent would have voted for the Peace and Democracy Party (Konda, 2014). It can be seen that the crowd that was in the park during the protests consisted of various factions and affiliations as well as no affiliations at all.

According to Konda's research, the various points at which people decided to attend the protests were as follows:

- When I heard about the Taksim project (10.2 percent)
- When they tried to uproot the trees (19.0 percent)
- When I saw the police violence (49.1 percent)
- When I heard the comments of Prime Minister Erdoğan (14.2 percent)
- When I saw the ambience in Taksim (4.3 percent). (Konda, 2014)

From the statistics above, it can be seen that 29.2 percent of the participants were motivated by protecting the existence of the park as it was: That is, 10.2 percent of the participants were moved to attend the protests because they opposed the demolition of the public park and the building of a shopping mall in its stead, and 19 percent were moved to attend the protests because they wanted to protect the trees in this public park—a space of leisure and a potential space of appearance in the biggest square of Istanbul. According to Konda's research, when the participants of the protests were asked about how they perceived themselves, 61.9 percent saw themselves as members of a group that faces infringement of human rights. When asked why they were in the park and what their demands were, 34.1 percent of the participants said they were there for freedom, with 18.4 percent saying they were against the infringement of rights and that they were demanding rights, 9.7 percent saying they were against dictatorship and suppression, 9.5 percent saying they demanded the resignation of the government, 8 percent saying they were for democracy and peace, 6.2 percent saying they were against police violence, 6.1 percent saying they were there for support and solidarity, 4.6 percent saying they were against uprooting the trees and building the artillery barracks, 3.2 percent supporting revolution against the system, and 2.5 percent saying they were there for their voices to be heard.

As can be seen from Konda's survey, there was no common agreement among the participants as to what signified the decisive moment to participate in the protests. What is interesting, though, is that close to half of them decided to participate when they saw the use of violence by the police. Consistent with Konda's research, almost half of my respondents also claimed that they went to the park after they saw social media posts alerting them to police violence.[16] Most of the respondents claimed that after they saw the police violence, they felt responsible to support the movement and the people in the park, and they felt responsible to inform others about what was going on. One respondent said, "I believe we all had to contribute somehow; this was a necessity to show the fact that the political dimension of Turkey had to understand that we exist." Respondents claimed that the more the police were seen to use violence, the more people were motivated to go to the park and even clash with the police. One respondent said,

> the state lied to the people, they said they were not going to attack, and they still did. When that happened, people came to Taksim with a bigger movement,

> when the municipal police burned the tents of people and were spraying water on people … these people are calmly sitting in the park trying to protect it … we wanted to protect the park, they attacked even though they said they were not going to, they said remove the barricades, we did remove them, they still attacked, I haven't seen anything like this in my entire life.

For those who saw all this police violence, it would seem to have been particularly dangerous to pour out onto the streets, since there was evidently a very strong risk of physical harm. Police violence, according to the respondents, proved to be the physical repercussion of the antidemocratic acts of the government up to the time that the Gezi Park protests occurred. The police violence conveyed to these people that the world as they wanted to maintain it was fragile, not only because of the threats against its material form but also because of violent attacks on its common context as a common world.

As mentioned earlier, during the Gezi Park protests, social media platforms—especially Facebook and Twitter—played a major role both in offering a space for the dissemination of information (nationally and internationally) and in facilitating communication among protesters (Chrona & Bee, 2017). Below, I examine the role played by these platforms during the protests.

Social Media During the Gezi Park Protests

During the interviews, my respondents singled out the first night of the protests as a wake-up call for them. It was on that first night that some people were already going out to support the protesters because they had seen the police violence, but the events of the evening were not televised. One respondent said,

> In the evening of the first night of the protests, I came home and told my mother that I am going to the park, and my mom asked me why. I was like, don't you know all hell broke loose in Taksim? She doesn't know, because on CNN Turk[17] there is a documentary on penguins, on NTV[18] there is a sports program … all hell broke loose in the heart of Istanbul, but nobody knew, the people that are not outside or don't use social media were not informed about what was going on … I heard this a lot, I mean people said if we don't hear what is happening in the heart of Istanbul, it is very normal that we didn't hear anything that happened in Diyarbakır[19] for years. If they are hiding the war in the heart of Istanbul that is only two streets away from me, how did they hide the war in the east? Or they told it to me the way they want to tell it … because it was like that, I also wouldn't

> know if I was not involved with it, I also was born and raised in Istanbul, I didn't feel the pain of the matter physically or mentally, nothing bad happened to me, but the more people heard, the more people started being in a state of disbelief, you say to yourself how can a state lie to its citizens to this level, how can a state be a liar like that? People started recognizing that their state is a big liar.

The participants of the Gezi Park protests were aware of the increasing censorship and government intervention in the news, and on the first day of the protests, they were faced with the fact that they needed to share information; otherwise, it was not going to be heard.

My respondents singled out the social media platforms Twitter and Facebook as the two main digital environments that were influential before, during, and after the protests. Most of them perceived these digital environments as having primarily an instrumental role.

Different social media platforms were used for different purposes. Some respondents did not have Twitter accounts, for example, but they opened accounts just to be informed and updated about what was happening in and around the protest area. One respondent observed a direct relationship between the political realm and social media:

> Social media exploded with Gezi in Turkey especially in terms of the impact of tweets, during Gezi that exploded, I think we realized the importance of that and after Gezi we couldn't depart from Twitter, I believe Gezi became a realm that made social media into a media. Besides, if you look at the places where social media exists politically, I mean where social media is important, you see that they are places where the political realm is opaque, like Turkey.

This observation represents a shared belief among the respondents. According to the 2014 report on freedom of the press by the Freedom House,[20] in Turkey,

> Constitutional guarantees of freedom of the press and expression are only partially upheld in practice, undermined by restrictive provisions in the criminal code and the Anti-Terrorism Act. Turkey remained the world's leading jailer of journalists in 2013, with 40 behind bars as of December 1, according to the Committee to Protect Journalists. The press freedom climate deteriorated sharply during the year as journalists were harassed and assaulted while attempting to cover the Gezi Park protests that broke out in Istanbul in May, and dozens were fired or forced to resign in response to sympathetic coverage of the protesters' demands. … The firings highlighted the close relationship between the government and many media owners, and the formal and informal.
>
> (Freedom House, 2014, pp. 11–12)

According to Freedom House, the press freedom status in Turkey in 2017 is "not free." Besides sharing information on the national level, the Gezi Park protesters also reached out to the international community during the protests, and social media platforms were the main mediators of this information flow between the protesters and the international media. In one day, according to a study from New York University (Tucker, 2014), about two million tweets were sent with the hashtags #direngeziparki[21] and #OccupyGezi. #direngeziparki was the first hashtag used, with 950,000 tweets, followed by #OccupyGezi with 170,000 tweets and #geziparki with 50,000 tweets. According to the report, 90 percent of the tweets were sent within Turkey, and the reason behind this explosion of tweets was the lack of coverage of the news by mainstream media. There were also thousands of tweets with the hashtag #BugünTelevizyonlarıKapat[22] as a reaction to this lack of coverage (Tucker, 2014).

Social media platforms like Facebook, Twitter, and YouTube were extremely influential during the Gezi Park protests. The major role of social media was, at the beginning of the protests, to lead the transition from a protest by a small environmentalist group to a large-scale, sociopolitical movement. Kamil Demirhan (2014) argues for the functionality of Twitter and Facebook during the protests. He delineates the functions of Twitter as "coordination among people, the mobilization of people and resources, the cooperation of actors, the dissemination of information, self-organization, creating new spaces to communication and including many people in the political discussions" (Demirhan, 2014, p. 292). Demirhan's analysis is very productive for an understanding of the instrumental character of social media use in the protests. According to Demirhan (2014) and Ozkirimli (2014), social media functioned as a tool in the hands of people to make the mobilization of the protests more efficient.

During an interview at the time on a Turkish television channel called HaberTurk, Prime Minister Erdoğan said, "There is now a menace which is called Twitter … The best examples of lies can be found there. To me, social media is the worst menace to society" (Letsch, 2013). When asked what they thought about the use of social media during the protests, one respondent said:

> The media here can be controlled, but as you see, the international media cannot be taken under control, at least from within here—they might have some control according to their own context—hence, it is very important in terms of being heard as well, also in terms of encouraging people to participate in the protests.

However, even while they used social media platforms during the protests, there was also mistrust and a sense of suspicion among the respondents related to the truthfulness of the information being disseminated via social media. One respondent said,

> In the national news, these things don't take place. Social media becomes the only source of news, even though it includes a lot of misinformation and it is very open to provocation and speculation about a lot of stupid things. Your only source of information is social media; you sit at work and watch Twitter all day long.

As can be seen, social media platforms were perceived as the only reliable source of information among the protesters, while they acknowledged the fact that these platforms harbored a lot of misinformation and information contamination. There was a shared belief in a direct relationship between information and political action among the protesters. All the respondents claimed that one had to know what was happening in order to be able to make up one's mind whether to participate in the protests or not. Because this was a shared belief, information, and the speed of the spread of information, became vital before and during the protests. When asked if the protests would have happened if there had been no social media, one respondent said,

> I wouldn't even know, I mean, there is still a lot of police violence today, a lot of people died, the laborers were being beaten up every week but none of them made us go out to the street. Maybe there were more violent police interventions than the one in Gezi but we didn't go out in any of them because on that day, there was a very impressive live stream and following that, we all, after that impressive live stream we saw a picture of an alternative world like people are singing and reading books, we said see, one can protest like this too, it is not really necessary to walk on the streets carrying a banner, we should participate in this.

According to one respondent, however, the protests would still have occurred:

> It was like a magnet, there was a magnet on social media, everybody started walking, everybody was writing Gezi, updating their locations we passed this point, we passed that point, whatever, you were always updated. Later, journal websites like 140 Journals acted like news agencies and started streaming live, people started following one another and we found ourselves around Taksim.

As can be seen, while being the medium through which protesters shared information and communication, social media platforms were also recognized as catalysts for the spontaneous eruption of people onto the streets and toward Taksim.

As mentioned before, the respondents shared the belief that having access to information would bring about political action. It is true that developing political knowledge and forming a political opinion necessitate information. However, people are bombarded daily with more information than they can process, and a critical perspective is required toward the information that is provided; one has to be able to distinguish between truth and a blatant lie. According to Papacharissi, simply acquiring or having access to information is not sufficient for being a more informed citizen (Papacharissi, 2002), and there is no direct relationship between being more informed and being motivated toward political action. Political action, or participating in a political protest, is more about one's disposition toward the world, which is informed by one's own principles, history, and community—the neighborhood one lives in or simply the people around one. Murat Özbank, in his book *Gezi Ruhu ve Politik Teori*[23] (Özbank, 2013), provides an incisive introduction to understanding the virtual and physical influences of materiality in political action. He argues that the participants of the Gezi Park protests turned social media into a public media (Özbank, 2013). Social media, such as Twitter and Facebook, served as a virtual public space during the protests, and the common will to act and resist came into being because of the equal and plural political actors who deliberated politically in these virtual public spaces (Özbank, 2013). According to Özbank, the political actors of the Gezi Park protests carried their common will, which was composed in the virtual public space, into the city squares of real cities as well (Özbank, 2013).

During the protests, social media was important for sharing information among the protesters and providing information to the national and international community whose members were not physically at the location. As mentioned, the interviewees assumed a direct relationship between being informed and taking political action. Social media was not only a place for this information to be disseminated but also a catalyst in arousing the attention of the protesters. Social media, then, was influential in instigating the awakening of the citizens' attention through allowing access to information. What needs to be noted here is that the role of social media was only complementary to what Gezi Park itself was doing in the mediation of the common sense. When the Gezi protesters tried to understand what was happening in the world and referred to their common sense, access to that common sense was mediated by Gezi Park, but being informed about what was happening in the world was mediated by Facebook. In the following subsection, I work out the mechanics of this technological mediation of common sense.

Technological Mediation of Common Sense in Gezi Park

Now, I would like to go back to my starting argument in this chapter: Like the cultural end products of *homo faber*, the material end results of *homo faber*'s work are capable of stabilizing the appearance of *praxis* retroactively and influencing the appearance of further *praxis* proactively.

I believe that until now, I have managed to convey that Gezi Park and the surrounding Taksim area are associated with freedom and plurality in the minds of Turkish people. This is, to an extent, due to the urban transformation of Istanbul during the twentieth century, when Taksim Square and Gezi Park became a showcase for the new Republic of Turkey, emphasizing "freedom," "equality," and "plurality," and transforming the obedient subjects of the Ottoman Sultan into free citizens of the Turkish Republic. Even though freedom, plurality, and equality were not always present in society at the time of the creation of Gezi Park, *praxis* strove for these values in the foundation of the new republic. Thus, the common sense of the time when the park was created was amplified when the park's existence was endangered. After seeing the police violence, for example, one respondent claimed to feel an urgent need to go to the protests as a result:

> I have to go because this is unacceptable, I had to pour down onto the streets and I did, literally … there is a very strong sense of guilt attached to it [felt by those who did not go to the protests] when you go to work the next day.

It is appropriate to pause here for a moment and consider what happened to my respondents. In the biggest square of the city they lived in, there was excessive use of violence against a handful of peaceful protesters by the police. This first encounter with this event, which was brought to the attention of the respondents thanks to social media, disrupted their routine behavior because it appealed to them. Then, they were moved toward understanding what was happening and why it was happening. They wanted to understand what was happening because the world that they shared with their fellow citizens was under attack. Because they care for the world that they share with other people, this police violence awakened their attention. The continuity and the maintenance of the common world, as they knew it and as they wanted to maintain it, was under physical attack.

As the reader will also remember, this appeal to taste and being moved toward understanding is only a preliminary understanding. This initial or preliminary understanding of phenomena is carried out through a reference to common sense, which is the shared interpretative framework of people. After an appeal to common sense, people grasp the initial meaning of phenomena and have a preliminary understanding. This preliminary understanding, as the reader will remember, opens the room for interpretation, therefore inviting explicit understanding. Now, it is possible to analyze the Gezi Park protests in light of the hermeneutic lemniscate.

The hermeneutic lemniscate in Figure 3 shows the structure through which the technopolitical mediation of technologies can be analyzed as well, in addition to showing moral dynamism, which is its purpose for Kudina. The world becomes a politically meaningful context when *praxis* (action) appears in the world. This appearance of *praxis* in the world passes through all three parties involved: the human who refers to common sense, the technology that mediates common sense, and the world that is co-original with common sense. There are two types of arrows that can be observed in this hermeneutic lemniscate: from the human to the world through technology, and from the world back to the human through technology. To enable analysis of the role of technology in the process of "political meaningfulness," these arrows are divided into four smaller ones.

An event occurred in Gezi Park on May 31, when a handful of protesters were attacked by municipal police. This information was brought to other residents of the city thanks to social media. Because the people read about or saw pictures of this attack, they wanted to understand the reason for it, or they wanted a fuller

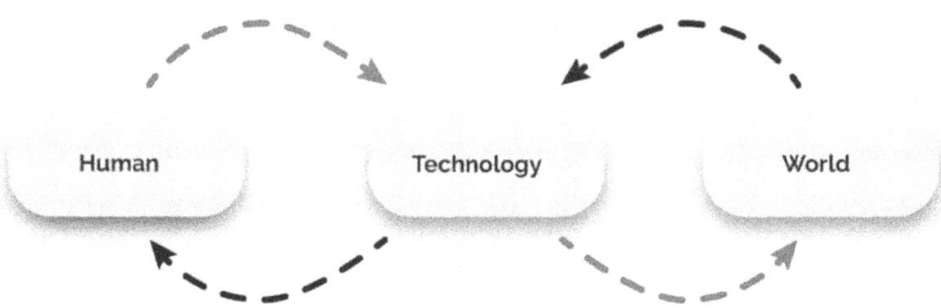

Figure 3 The Hermeneutic Lemniscate (the first move).

understanding of the event they had read or heard about. They saw a group of people protesting about something. At first, they did not know what this event was, but in order for them to understand what it was, they referred to their preliminary understanding—that is, to their common sense. Situated between their access to the world where this event occurred and their referral to their common sense was Gezi Park. From their common sense, they knew that the space occupied by Gezi Park is a space of freedom. They knew this because Gezi Park and the surrounding Taksim area had been appropriated by people before them (who contributed to this common sense that they share) as a public space for civic engagement. They knew of the plans by the municipality of Istanbul and the government of Turkey to demolish this park and build a shopping mall, which would be a replica of the Ottoman artillery barracks. They knew these plans would mean that this space of appearance—which hosted freedom, plurality, equality, and the free civic life of citizens—would be destroyed. Demolishing Gezi Park meant demolishing free citizens and creating a new form of subject, one that this time would obey not the Sultan but the government ruled by the AKP. When the citizens of Turkey saw this phenomenon, it attracted their attention, since Taksim means freedom in the minds of Turkish people. The intervention into the park meant an intervention into the freedom of people in Turkey. Gezi Park then amplified one aspect of its meaning in Turkish society as being a space of appearance (which can be seen in the diagram of the lemniscate in the move from the upper left curve toward the lower right curve).

From an appeal to the common sense and an articulation of this preliminary understanding, the citizens judged that the world as they knew it was in danger, and then they were moved toward action. Looking back at the hermeneutic lemniscate model, it can be seen that it proves helpful in understanding what a material and digital construct or technology does. Following the move from the upper left curve to the lower right curve, it can be seen that the human access to common sense and the common world was mediated by the technology, in this case the physical space of Gezi Park itself and social media such as Facebook and Twitter. Gezi Park amplified the common shared aspect of the world and drew these people together for its maintenance, not only because it was in danger but also because of what it means for the people of Turkey. Social media amplified this existing reality and provided access to the world through information. Gezi Park mediated common sense, as it is the material form of the cultural memory of Turkey. The Turkish people knew what Gezi Park represents because of their common sense. Following Arendt, we know that it is only through human appropriation of it that a space turns into a public space. Thus, this human appropriation, even if it is only temporary as in the example of Gezi Park, is not in vain (i.e., it has succeeded in achieving this transformation of space). Only when humans appropriate a space through their *praxis* does that space turn into a space of appearance.

Conclusion

In this chapter, based on Arendt's analysis of the role of the cultural end products of *poiesis* to stabilize the appearance of *praxis* retroactively and influence the further appearance of *praxis* proactively, I have demonstrated how this role extends to material products as well. I have shown, from the perspective of postphenomenology of technology, that things (artifacts, objects, and technologies) have a hermeneutic dimension. Because of this quality, they mediate human experiences and how people have access to their world—in other words, how people understand their world. In addition, following Kudina (2019), I have shown that human interpretation and sense-making are technologically mediated while acknowledging that the cultural and moral history of the person engaging with the technology in question is essential in this mediation.

Following the (post)phenomenological turn to empirically informed microlevel studies, I conducted an empirical investigation, which enabled me to examine my proposed extension of Arendt's political hermeneutics to materiality using a real example. To this end, I examined the active role of Istanbul's Gezi Park in initiating the Gezi Park protests and forming a heterogeneous public during the protests. By providing extracts from the interviews that I conducted with the protest participants, I have shown that Gezi Park has a place in the common sense of the people of Turkey. I demonstrated that Gezi Park is a political space for the Turkish people due to its importance in the common cultural context of the people. I showed that the people in Turkey resisted the threatened demolition of Gezi Park because of its important place in their common cultural framework.

In this chapter, I have shown that Arendt's political hermeneutics, once extended to materiality, allows us to grasp the political quality of technologies in its fullest sense. Thanks to this extension, we can argue that, on the one hand, technologies are a representation of our common interpretative frameworks, and, on the other hand, they inspire political action because of this fact. However, as I show in the next chapter, this is only the first step in this book's quest to unpack the political role of technologies.

Until this point, I have shown that a material construct, Gezi Park, played an active role in instigating political action, namely the Gezi Park protests. However, I have not investigated the role this material space played in generating a heterogeneous public by mediating the interactions of the people. In Chapter 4, I take on this task and investigate the role of technology in informing the way in which political communities are formed by mediating our political interactions.

Notes

1 When I explained Arendt's emphasis on experience, I mentioned that experience is grounded in the world. See Chapter 2, pp. 56–61.

2 See Chapter 2, p. 57.

3 As I formulated earlier, in Chapter 2, Arendt's use of common sense as a *sixth sense* corresponds to this microperception/macroperception scheme that Don Ihde (1993) formulates in his postphenomenology. Thus, it should be noted here that the macroperception is closely related to how Arendt formulates common sense as the sixth sense that fits our sensory perceptions into a common context and thus allows us to make sense of the world.

4 Here, Verbeek does not refer to Arendt's use of the term *praxis*. His use of the term strictly describes how humans are present in the world (Verbeek, 2006). Verbeek indicates, "From a phenomenological point of view, praxis and existence are the mirror images of perception and experience. Whereas perception consists in the way the world is present for humans, praxis can be seen as the way humans are present in their world" (Verbeek, 2006, p. 58).

5 Don Ihde (1990) considers all material artifice as a technology, from clothing to culinary technologies. He indicates that his concept of technology is as "broad as possible while still retaining an emphasis upon its materiality" (Ihde, 1990, p. 21). Ihde argues: "virtually every area of praxis implicates a technology. From burial to birth to eating and working, the use of artifacts embedded in a patterned praxis demarcates the human within his or her world. To reverse the usual equation, the technological form of life is part and parcel of culture, just as culture in the human sense inevitably implies technologies" (Ihde, 1990, p. 20). Following Ihde, it is not too far-fetched to consider the material artifact Gezi Park as a technology as well.

6 My explanation.

7 My explanation.

8 Gezi Park is only one material infrastructure—the one that I pay attention to—within many other structures in Prost's master plan of Istanbul. The whole master plan consisted of "the division of the city into industrial, commercial, residential and recreational zones and the establishment of an effective transportation system" (Gül, 2009, p. 98). It is not within the scope of this book to analyze the ideological transformation of the city of Istanbul in line with the secular, Kemalist republic. For more on this, see Gül (2009).

9 Translated from Turkish. Original: "Bugün 'Avrupalılaştırma, nezihleştirme, sekülerleştirme, vatandaşlaştırma' üzerinden de okunabilecek geometrik düzen, sırayla dikilmiş ağaçlar, muntazam çiçek tarhları, havuz ve heykeller Taksim'de mimari kurgudan çok daha fazlasına işaret ederler" (Akpınar & Şoher, 2013, p. 5).

10 Translated from Turkish. Original "Prost, karma bir toplum için kamusal mekanlar ile master plandaki önerilerinin radikal sosyal ve kültürel donuşumu desteklediğini yineler" (Akpınar & Şoher, 2013, p. 5).

11 In this work, Karen Till investigates the intricate relationship between social memory, national identity, and places such as museums and memorial spaces. For more on this, see Till (2005).

12 According to the literature, the headscarf is perceived in Turkey as a symbol of resistance against the Kemalist reforms that aimed at the separation of state and religion, which was the basis of the new Republic of Turkey (Lewis, 1968).

13 It is not within the scope of this book to discuss in depth the conflict between Islam and Kemalism in Turkey. However, I consider it useful at least to mention that the Islamism of the AKP government and the support of its "conservative democracy" have deep historical roots in Turkey. For more on this, see Ahmad (1993).

14 For more on this, see Baykan and Hatuka (2010).

15 Gezi Park Protests.

16 The police violence against the protesters on May 31 was not covered by any mainstream media in Turkey.

17 News media channels in Turkey.

18 News media channels in Turkey.

19 City in the southeast of Turkey.

20 Freedom House (2014). An independent organization heavily funded by the US government that conducts research and advocates the expansion of freedom and democracy around the world: https://freedomhouse.org.

21 Resist Gezi Park (my translation).

22 Turn off the television today (my translation).

23 Gezi Spirit and Political Theory (my translation).

4

Technological Mediation of Intersubjectivity

In the previous chapter, through the example of the Gezi Park protests, I showed that this public park mediated the shared political interpretative framework of the protesters through mediating the common meaning given to the park as a space of freedom. To unravel the role of the technology—namely Gezi Park, in this case—in the political community that was formed during the protests, a next step has to be taken. This next step necessitates an investigation into the structure of a political community (the Gezi Commune) that defined itself as "unified" but not homogeneous (Acar & Uluğ, 2014). Furthermore, it necessitates an investigation into how the design of the park, and the relationship between the park and the protesters, enabled this political community. In this chapter, I argue that technologies mediate the way in which our political communities are constituted by mediating our political interactions. I develop and prove this argument in five sections.

In the first section of the chapter, I focus on demonstrating the interconnected relationship between political subjectivity and intersubjectivity. First, I unpack Arendt's definition of the political subject and show that Arendt conceptualizes the political subject as "authentic," a description she appropriates from Heidegger's notion of *Dasein* as authentic human existence (Villa, 1995). This, I argue, allows Arendt to explain political subjectivity as a process that is influenced by the conditions surrounding the person. Second, I show that while Heidegger presents being in a community as a danger for the manifestation of the authentic and free self, Arendt shows that community is rather a condition for this manifestation rather than a threat to it. Third, following Arendt, I show that the political community one is engaged in has a conditioning force upon the political subject. This allows me to argue that in order to understand the role of technology in political interactions among political subjects, it is necessary to consider the type of political community that is fostered by the material space that engulfs that community.

Having presented these distinctions, in the second section of the chapter, I focus on demonstrating the role of technology in this interplay between acting agents and in the intersubjective relationships between the acting agents. Through a thorough analysis of the existing literature and the interviews I conducted with the participants of the Gezi Park protests, I show that the organization of the protesters within the park during the protests constituted a specific interpretation of sharing a common space.

In the third section of the chapter, I show that Arendt's definition of public space (which necessitates plurality of perspectives and a stable material space) is intrinsically related to the concept of multistability, through which postphenomenology of technology analyzes artifacts. I argue that the more multistable a public space is, the more it invites multiple actors to visibly manifest their specific interpretations and share a common space in their uniqueness and distinctness. I illustrate that the multistability of the space that hosts a community is an essential condition for the authenticity of the community (and eventually of the individual). I show that a community can be considered authentic when its members create new meaning through the appropriation of an artifact and allow differing interpretations to coexist while denying any "stabilized meaning" of the artifact in question. By contrast, the inauthentic community would embrace the "stabilized meaning" of an artifact while suppressing alternative interpretations and appropriations of the artifact in question. I show that the organization of the protesters within the park during the protests embodied a specific interpretation of sharing a common space. This interpretation, however, was the process of technological mediation of intersubjectivity: the interplay between humans and the material design of the park opened new realms of action within the world through fostering engagement among diverse subjectivities.

In the fourth section of the chapter, I demonstrate technopolitical mediation of technologies as a two-step process. I show that in order to observe technological mediation of intersubjectivity, it is first necessary to observe technological mediation of common sense. Even if mediation of common sense is bound to occur for anyone who engages with a technology, how the interplay between the subject and the intersubjective community will play out is to a certain extent dependent on the multistable design of the technology in question. In addition, the existence of a political community that appropriates a space as a public space is also dependent on the context of the mediation of common sense. I repurpose Kudina's (2019) hermeneutic lemniscate to show that the material space where people meet mediates both the way in which a community is established and the way in which one becomes a part of a community.

In the fifth section of the chapter, I show that technopolitical mediation is not limited to physical material spaces but also applies to digital spaces. I apply the framework I have developed to Facebook and its role during the Gezi Park protests, since it is the digital platform that was singled out as influential during the protests by my respondents.

Political Subjectivity and Intersubjectivity

As mentioned before, the reason for Arendt to portray political action as *praxis* is to protect political action from being reduced to an expression of one's identity. Arendt perceives action as showing the care of the acting agent for the world, and in that activity, the agent discloses themselves as well. In addition, Arendt argues that the way in which humans act and react to things can change in relation to the events, people, community, and the institutional and cultural structure surrounding them. *Praxis*, in this sense, is constitutive of the individual agent as a political subject in its performativity. The acting agent simultaneously discloses and discovers who they are while carrying out *praxis*. Therefore, there is no one stable identity of the political subject: It is a process of *becoming* rather than *being*. Honig defines Arendt's conceptualization of self as follows:

> Prior to or apart from action, the self is fragmented, dis-continuous, indistinct, and most certainly uninteresting. A life-sustaining, psychologically determined, trivial, and imitable biological creature in the private realm, this self attains identity—becomes a "who"—by acting in the public realm in concert with others. In so doing, it forsakes "what" it is, the roles and features that define (and even determine) it in the private realm.
>
> (Honig, 1993, pp. 79–80)

Honig's elaboration reveals two dimensions of the self that are implicit in Arendt's configuration: the "what" and the "who" of someone. While the "what" of the actor is determined by external roles like belonging to a family (being a mother or daughter), the "who" of the actor is the end result of a gradual process in which the actor attains different roles every time they become visible in the world through portraying a concern for the world. In that sense, the "who" of someone is responsive to the events unfolding within the world and to the people with whom they share the world (Arendt, 1998). The "who" of someone becomes visible when they act in the world, driven by care for the world (*amor mundi*).[1] However, this care—the form and the context of it—is never static because of the ephemeral character of *praxis*. The recognition of care and responsibility toward the world changes the actor's relation to the events, people, community, and the institutional and cultural structure surrounding them. This can result in portraying different aspects of the same agent as a "who."

For Arendt, with every new response to the events of the world, with every act of care and concern for the world, a new "who" appears accordingly (Arendt, 1998). This is why, for Arendt, every action is a new beginning, and "who" someone is can change according to the event, the response, and how others with whom the actor engages are going to perceive the actor and the action itself (Arendt, 1998). As action (*praxis*) is spontaneous, unpredictable, and a temporal activity, it allows a "who" to appear in its execution, and that appearance accumulates in the life story of the actor. According to Arendt,

> who somebody is or was we can know only by knowing the story of which he is himself the hero … everything else we know of him, including the work he may have produced and left behind, tells us only what he is or was.
>
> (Arendt, 1998, p. 186)

Arendt, in this quotation, asserts that the "who" of someone is a story that we see and hear from others who saw and heard that person. There is no essence of a person,[2] but rather there is the story: the narrative that person leaves behind within the world. Arendt argues:

> the stories, the results of action and speech, reveal an agent, but this agent is not an author or producer. Somebody began it and is its subject in the twofold sense of the word, namely, its actor and sufferer, but nobody is its author.
>
> (Arendt, 1998, p. 184)

For Arendt, the definition of a "who" does not refer to a human essence or human nature. The end appearance of the "who" is outside of the control of the acting agent; it is left to the discretion of the community in which that agent engages. In that sense, the unique life story—that is, the "who" of a person—might not even be what that person intended. This shows that the "whoness" of someone is inextricably linked to intersubjectivity. Who a person is, and who they are going to be, is not only dependent on that person themselves. Humans can show how they care for the world and take responsibility for the world through their actions and speeches. Someone's "whoness" is also dependent on the community they become *visible* to. In order to be able to explain the inextricably linked character of "whoness" with intersubjectivity, in the next subsection, I turn to Arendt's reinterpretation of Heidegger's *Dasein* as an authentic mode of existence as a political subject.

Dasein: Authentic or Inauthentic Modes of Human Existence

For Arendt, a political subject is one who is able to create meaning in the world. Only the creation of meaning allows one to be on the path of authenticity. The question then becomes, how does Arendt decide what is meaningful and thus define a person as an authentic agent? Here, a detour to Heidegger's distinction between authentic and inauthentic existence as disclosedness is necessary since Heidegger's distinction forms the basis for Arendt's configuration of the "who" of someone as a politically relevant subject. However, Arendt moves away from Heidegger when she characterizes the gradual path to authenticity as being a move toward the world, rather than a move toward the self, as Heidegger defines it. First, let me explain how Heidegger defines authenticity.

For Heidegger, *Dasein* is authentic when it is transcending the conditions that already exist in the world through its actions, and *Dasein* is inauthentic when it is absorbed in everydayness. Let me clarify this further.

As mentioned before, *Dasein*, according to Heidegger, is a with-world (*Mitwelt*), which means that there is no human existence that is outside of the world (Villa, 1995). This means that *Dasein* is always in the world; in other words, *Dasein* is being-in-the-world[3] (Villa, 1995). This world that we live in shapes and changes us, because both other humans with whom we share the world and the material objects that are in the world are influential in the way we exist. *Dasein* is constantly disclosing the world, but how *Dasein* is in the world is defined by its mode of being in the world: either as authentic or inauthentic existence.

For Heidegger, *Dasein* has a "built-in tendency to 'forget' its disclosive or projective character" (Villa, 1995, p. 127). This means that, in its everyday mode—in its everydayness, so to speak—*Dasein* moves away from realizing itself, and it becomes busy with the things, people, and concerns that it encounters (Villa, 1995, p. 128). In its everydayness, *Dasein* is inauthentic, according to Heidegger, because it fits itself to others. When *Dasein* is in its everyday mode, it is a "being-in-the-world" of "things already produced either by human beings that existed in the past or by those existing in the present"

(Mansbach, 1991, p. 75). This means that the world already has a certain form and structure of meaning that was created by others. In its everydayness, *Dasein* is not concerned with meaning creation, showing its unique stance about the matters of the world, since meaning has already been created, and what those matters mean has already been defined (Mansbach, 1991).

The inauthentic existence is where *Dasein* finds itself in a world that already has meaning and discloses itself, understands itself, and projects into the future within the existing framework of meaning that was prescribed by others (Mansbach, 1991, p. 75). In that sense, *Dasein* even discloses itself within this framework of meaning, and therefore, there is no self-understanding or self-realization that comes from *Dasein* itself. In this inauthentic existence, as Heidegger says, "everyone is the other, and no one is himself" (Heidegger et al., 1962, p. 165). In the mode of inauthentic existence, *Dasein* is under the domination of others that dictate *Dasein*'s self (Mansbach, 1991, p. 75). Our everyday practices—the things we talk about and the way we do things—seem to provide "a full, adequate, and complete grasp of existence" (Villa, 1995, p. 128) for *Dasein*. It seems as if there is nothing to disclose, because everydayness has an explanation for everything there is.

Heidegger also describes the everydayness as disclosing the world only in means–ends terms—in other words, perceiving everything and everyone as ready to be manipulated and exploited. This manner of perceiving the world does not enable one to question the meaning of the events and relationships around oneself but only their efficacy and usefulness. Therefore, *Dasein* loses its own capacity to grasp who the individual is. Everything that is distinctive about *Dasein* becomes lost and absorbed within others. Here, a category of *Dasein* becomes apparent: the "they" (*Das Man*). The "they," as Mansbach puts it,

> explains human existence uniformly. The "they" is neither the totality of individuals, nor a universal subject which the human being is part of. It is the anonymous crowd to which every human being belongs, any member of which can be substituted for or represented by any other.
>
> (Mansbach, 1991, p. 75)

It is the "they" that Heidegger calls the public. The public moves *Dasein* away from itself and its world, because the public only objectifies the world and itself. Heidegger asserts,

> Idle talk and the way things have been publicly interpreted… holds *Dasein* fast in its fallenness. Idle talk and ambiguity, having seen everything, having

> understood everything, develop the supposition that *Dasein*'s disclosedness, which is so available and so prevalent, can guarantee to *Dasein* that all the possibilities of its Being will be secure, genuine, and full. Through the self-certainty and decidedness of the "they," it gets spread abroad increasingly that there is no need of authentic understanding or the state of mind that goes with it. The supposition of the "they," that one is leading and sustaining a full and genuine "life," brings *Dasein* a *tranquillity*, for which everything is "in the best of order" and all doors are open.
>
> (Heidegger et al., 1962, pp. 221–2)

Being absorbed in the public—in other words, inauthentic existence—is predominant in the self, according to Heidegger. As mentioned above, the instrumental way of perceiving the world and others causes *Dasein* to lose its capacity to make connections between different people and different objects within the world. Therefore, *Dasein* loses its capacity to create meaning within the world. However, there is a choice. For Heidegger, once *Dasein* is faced with its own mortality, when death is disclosed, *Dasein* is forced to "confront the lack of meaning of things and others" (Mansbach, 1991, p. 77). This confrontation enables *Dasein* to grasp its being for the sake of itself (Villa, 1995, p. 131). This means that *Dasein* "does not settle for the security, familiarity of the everyday world" (Villa, 1995, p. 131) but transcends these conditions by rejecting the world as already having all the meaning that it is possible for it to have (Villa, 1995).

This is the moment when *Dasein* is able to transcend the everydayness and disclose the world. *Dasein* as authentic existence appropriates what is already in the world and provides a new perspective and new meaning within the world. It strives to disclose and to "uncover the new through its creative or originary spontaneity" (Villa, 1995, p. 132). For Heidegger, "care" and authentic existence are strongly connected with one another. Without care, it is impossible to reach an authentic mode of existence, since Heidegger argues that

> Dasein's Being reveals itself as *care*. If we are to work out this basic existential phenomenon we must distinguish it from phenomena which might be proximally identified with care, such as will, wish, addiction and urge. Care cannot be derived from these, since they themselves are founded upon it.
>
> (Heidegger et al., 1962, p. 227)

This shows that the essence of *Dasein* reveals itself through care; therefore, it is care that leads the way for authenticity, according to Heidegger. Inwood (1999) explains that Heidegger uses three different words in his description of care: *Sorge, Besorgen,* and *Fürsorge* (Inwood, 1999, p. 35). Heidegger defines care as *Sorge* when he refers to the apprehension and anxiety that arise from fear of the future; the word refers to both the internal condition and an external cause. *Besorgen* means obtaining something for oneself or someone else or making something available (Inwood, 1999, p. 35). *Fürsorge* means actively caring for someone in need (Inwood, 1999, p. 35). The concepts differ in that *Sorge* relates to existence itself, *Besorgen* is about activities in the world, and *Fürsorge* is about being with others. *Sorge* is the predominant mode of the three while being inseparable from the other two. According to Heidegger, there is a temporal aspect to care (Inwood, 1999, p. 36). *Sorge* is concerned with one's own past actions; *Besorgen* is concerned with the present through engaging with others; and *Fürsorge,* by being concerned with others (and not things), thinks of the future (Inwood, 1999, p. 36). In that sense, the three modes of care correspond to past, present, and future, and "temporality reveals itself to be the sense of authentic care" (Inwood, 1999, p. 37). Through the acknowledgment of care, we are able to *let things be.* In that way, we remove ourselves from the structures of dominating or being dominated, and we take a more open approach to the world, since being-in-the-world is essentially care (Inwood, 1999). *Dasein* moves away from its everydayness through disclosing, which "is a mode of activity and understanding that breathes new life into the familiar" (Villa, 1995, p. 132).

Dasein's move away from everydayness does not mean it is moving away from the world; rather, it means that *Dasein* opens a different perspective and a different range of projections within the world. As Villa argues,

> If authenticity means anything, it means a willingness to put oneself at risk in the opening of new possibilities, the willingness to abandon the security and tranquility of the ground provided by everydayness.
>
> (Villa, 1995, p. 133)

Authentic existence, then, is breaking the habitual way of being, erecting something new, and appropriating the world in a new way. However, for Heidegger, the authentic existence is a solitary endeavor.

As overwhelming as everydayness is, *Dasein* can still escape from this mode and be open to its own future when it is faced with death, which is related to an intense feeling of *Angst* (anxiety), according to Heidegger. In an intense moment of *Angst* that is generated by facing death,[4] *Dasein* is presented with a "moment of vision" (Heidegger et al., 1962, p. 387). According to Heidegger, this moment of vision

> must be understood in the active sense as an ecstasis. It means the resolute rapture with which Dasein is carried away to whatever possibilities and circumstances are encountered in the Situation as possible objects of concern, but a rapture which is *held* in resoluteness.
>
> (Heidegger et al., 1962, p. 387)

The moment of vision allows *Dasein* to transcend the condition of inauthenticity (everydayness) and brings it back to the present moment, allowing it to open up. As Inwood describes it,

> Heidegger associates resoluteness with being opened up: "Resoluteness [Entschlossenheit] is a distinctive mode of Dasein's disclosedness [Erschlossenheit]." *Entschlossenheit* is "the *specifically* undertaken self-opening of Dasein *for* the open" (G, 59/81). He also uses *beschlossen*, … "to include" and also "to decide," i.e. to conclude one's thoughts on the matter.
>
> (Inwood, 1999, pp. 186–7)

According to Heidegger, "when resolute, *Dasein* has brought itself back from falling, and done so precisely in order to be more authentically 'there' in the 'moment of vision' as regards to the situation which has been disclosed" (Heidegger et al., 1962, p. 376). Resoluteness is the response to the moment and the situation, and it is a mode of contemplation that is an articulation of the call of conscience. This conscience, as Heidegger defines it,

> gives us something to understand; it *discloses*. This disclosedness, as a basic state of that entity which we ourselves are, is constituted by state-of-mind, understanding, falling and discourse. If we analyze conscience more penetratingly, it is revealed as a call. Calling is a mode of *discourse*. The call of conscience has the character of an appeal to Dasein by calling it to its ownmost potentiality-for-Being-its-Self.
>
> (Heidegger et al., 1962, p. 314)

However, this call of conscience requires a hearing of conscience, which can only be reached through resoluteness (Heidegger et al., 1962, p. 134). That is why, for Heidegger, "resoluteness … represents an authentic disclosedness of Dasein" (Heidegger et al., 1962, p. 384).

As Mansbach explains, Heidegger's distinction between authentic and inauthentic existence should be understood as "authentic and inauthentic ways of *being-a-Self* ... this suggests that human existence, and all human behaviour, authentic or inauthentic, must be interpreted as the making of the self" (Mansbach, 1991, p. 80). In this subsection, I have described Heidegger's *Dasein* as authentic existence since Arendt is heavily influenced by *Dasein* in her understanding of the political subject. However, Arendt strongly disagrees with Heidegger's understanding of the public that makes *Dasein* move away from itself. Rather, Arendt depicts the public as the only realm where *Dasein* can be authentic.

Arendt's Reinterpretation of Dasein as Authentic Political Subject

One of the main concerns related to Heidegger's *Dasein*, for Arendt, is that *Dasein* asks itself the question "Who am I?" and answers this question by itself as well. Arendt disagrees with Heidegger here, since, according to her, if authenticity depends on *Dasein*'s meaning-creation capacity, then *Dasein* cannot create meaning in the world by itself and for itself only. In other words, this meaning creation cannot be executed without the existence of others. This means that she rejects the lonely, resolute authenticity of *Dasein*. The reason for this rejection stems from how Heidegger situates freedom.

According to Nichols, Heidegger perceives a strong connection between *Angst* and freedom (Nichols, 2000). As already mentioned, *Sorge*, worrying about the *Dasein*, only comes to the fore when existence is confronted with fear, the uncanny nothingness. But this nothingness, as Nichols explains, is

> *Angst*, (the oppressive nature of the *nothingness* encountered in Dasein's authentic potentiality-for-being-in-the-world; the fact that Dasein might *not* actualize its *being*—i.e., might not *be*—authentically; SZ 187/175), discloses Dasein's fundamental possibility of "*being free for* the freedom of choosing and grasping itself" (emphasis in original; SZ 188/176). And this is its ownmost potentiality of being.
>
> (Nichols, 2000, p. 3)

By connecting freedom to *Angst*, Heidegger is able to situate "not-being-at-home with one's self" (Nichols, 2000, p. 3) as revealing *Dasein* as always already ahead of itself (which is existentiality—also known as *Sorge*). As Nichols explains,

> In Dasein's "being-ahead-of-itself" (in *Sorge*), beings are "freed up" to be the kind of beings which they potentially "are" (in our prethematic use of them), and

> at the same time (*gleichursprünglich*) Dasein is thrown into its potentiality for becoming an authentic self. (SZ 193/180)
>
> (Nichols, 2000, p. 3)

This shows that for Heidegger, freedom *is Dasein*'s potentiality for being authentic. The self-actualization of *Dasein* is only possible through *Sorge*, a concern with one's self. The "realization" (Nichols, 2000, p. 3) of this concrete existential potentiality can only be attained through resoluteness (Nichols, 2000). In other words, Heidegger argues that freedom is not "absence of constraint with respect to what we can or cannot do. Nor is it, on the other hand, mere readiness for what is required or necessary" (Heidegger, 1997, p. 128). By denoting here what freedom is not, Heidegger defines freedom as "the engagement in the disclosure of being as such" (Heidegger, 1997, p. 128). Freedom is the mode of being that makes possible the will as a faculty of choice or decision (Villa, 1995, p. 126). Villa explains:

> The kind of freedom Heidegger is pointing to is neither "mere absence of constraint" nor "readiness for what is required and necessary" (an ought or duty): it is, rather, a freedom prior to negative and positive freedom, a freedom that is the condition of possibility for both. This freedom—the ontological freedom of the being whose being is thrown projection—is the freedom of the "open region," the freedom found in the "engagement in the disclosure of beings as such." It is the freedom of disclosedness: the freedom of an "open comportment" toward the world that both animals and things lack, and that the will (our organ of choice or purposiveness) covers over; it is the freedom of the "there," a freedom *for* the world.
>
> (Villa, 1995, p. 126)

Thus, freedom is having the possibility of willing. This formulation of freedom is highly problematic for Arendt. If freedom is *Dasein*'s potentiality for being authentic, this would mean that freedom is an inner-worldly affair.

As mentioned before, Arendt considers *Dasein* as authentic existence to be the only authentic and thus politically relevant subject. However, she does not agree with how Heidegger situates freedom only with regard to the resolute being. As demonstrated in Chapter 1, Arendt does not position freedom as the property of a subject. Her main concern in formulating freedom as a worldly phenomenon is her conviction that one cannot be free if one's freedom is not actualized within the world. Thus, inner freedom (as Heidegger would have it) or freedom of the will, for Arendt, are irrelevant to what she perceives real (political) freedom to be and its relation to authenticity.

Arendt contests the surface and depth in subjectivity by stating, "our common conviction that what is inside ourselves, our 'inner life,' is more relevant to what we 'are' than what appears on the outside is an illusion" (Arendt, 1981a, p. 30). However, only the world of appearances is the world of others. In the world of appearances, one shows oneself through deeds and actions, whereas in the world of the mind—that is, the world of self—one reveals oneself through the medium of thinking. As Arendt states, "appearance is by no means the outward manifestation of an inner disposition; if it were, we probably would all act and speak alike" (Arendt, 1981a, p. 34). Thus, for Arendt, freedom cannot be limited to the world of self, as freedom is a worldly phenomenon, not an inner-worldly one. According to Arendt, if we are to accept that freedom is a worldly phenomenon, then Heidegger's *Dasein* as authentic existence cannot be free, since that requires a move toward the self.

While appreciating Heidegger's *Dasein*, Arendt focuses on the features that make up the world in *Dasein*'s being-in-the-world, since freedom could only be found in *Dasein's* relation to the world, not in *Dasein*'s move away from the world. Arendt, by defining the being-with of *Dasein* as being-with-others, argues that the authenticity of someone can only come to the fore within a web of relationships that person establishes with others.

As already mentioned, according to Heidegger, *Dasein* is with others when it is in public, and when *Dasein* is in public, the self can only be the inauthentic self. This, for Arendt, is an incorrect description of the public. Maintaining that being-with-others can also take the form of the "they," Arendt formulates this way of being-with-others as "the social."[5] As mentioned in Chapter 2, Arendt formulates the social as a realm that is blurred, where there is no distinction between private and public (Arendt, 1998). Just as Heidegger does when discussing *Dasein*'s relationship to the "they," when Arendt defines the realm of the social, she talks about the need for and expectation of uniformity, being driven by necessity, and exchanging conversations in the form of idle talk.

Being-with-others in public, for Arendt, is the only way the self can be authentic. Arendt's notion of the political sphere seeks to demonstrate what *meaningful* existence entails. For Arendt, this is caring for the shared common world and orienting oneself toward a common interest that is not directly related to the self but external to it, allowing the authentic self to come to the fore (Kelz, 2016).

According to Arendt, action is a risky endeavor that requires courage, since it implies leaving the safety of the private sphere, where one is concerned with one's survival, and instead having one's opinions criticized and judged by others (Arendt, 1998). This act of leaving one's private space in order to be seen and heard by others requires a willingness to act, and this willingness to act has its basis in *amor mundi*. As I mentioned in Chapter 1, this is the basis for political freedom: the capacity to turn the inner "I-will" into "I-can" and exert oneself to move from the secure, invisible space of the private to the ambiguous, visible space of the public. Enacting this form of freedom requires courage on the part of the acting agent, since now, every act one does and every speech one utters will be open to public scrutiny. Arendt maintains that because we insert ourselves into the public, every act that is visible to others portrays the courage of the actor, who brings something new into this world through their interpretation of how the world appears to them or their unique perspective on the matters of the world. In other words, for Arendt, courage means taking a stand publicly and showing that stance to others in the public realm. In that sense, for Arendt, authenticity has a direct relation to freedom, and it is situated in public. This is how humans gain visibility, and that implies taking responsibility for one's actions, but the outcome of these actions is never clear beforehand. Because action is always an interaction with others who have different views, it depends on the existence of plurality. Only within the public do humans have access to different selves. All these different selves are unique and distinctive in their own access to and interpretation of the world. According to Arendt, only in interaction with these diverse accesses to the world can people be open to and able to grasp the reality of appearances. An individual affirms the validity of their perspective by taking a stance on matters; they let stances that are in contradiction with their own perspective be, and they learn to exist and coexist.

In this interaction with others, one expresses who one is, which also means that one acts and becomes who one is during the interaction. This is also directly related to Arendt's rejection of *Dasein*'s orientedness toward death. For Arendt, rather than being thrown into the world, we "insert ourselves into the world" (Arendt, 1998, p. 176). By likening human action to natality in describing it as the "second birth" (Arendt, 1998), Arendt shows that only through interaction with others can a person realize their potential for authenticity; this cannot be realized by confronting death. There is no stable self, and even if there were, we would not have access to it. What we have access to is, as Arendt states, "natural and unavoidable illusions of our sense apparatus .... an 'inside self,' if it exists at all, never appears to either the inner or the outward sense" (Arendt, 1981a, pp. 38–9). That is why, for Arendt, who we are is a constant work in progress, since we portray different facets of ourselves every time we interact with others and act within the world.

Arendt's insistence on plurality, rather than singularity, in order to reach authenticity provides a political twist that emphasizes visibility in public rather than solitude in isolation. Arendt's association of action (*praxis*) with natality also provides the basis for her rejection of a Heideggerian understanding of a human essence. Rather, for Arendt, humans are born and reborn anew every time they act, and each act does not provide any appearance or information about the person's inner life but only about that person's act. This stems from Arendt's phenomenological conviction that every act needs to be perceived and judged according to its temporality, its spatiality, and the intersubjective realm in which that act is being carried out. All these acts are unique stories of the acting agent—glimpses of who they are. The whole story of someone is alien even to the person themself. We, as the audience, will see who someone is only after that person's death, when they are no longer an acting agent. This allows Arendt to move away from a conception of human essence that would enable us to configure how someone would act in a given situation. Instead, the agent is allowed room to be free and construct a "who" of themself as a fluid concept that changes and takes shape in relation to the time when the agent is living, the environment that the agent is in, and the relationships that the agent builds around themself. Arendt also points out that this public appearance to create a new story every time one acts requires courage, because with each act, one shows one's unique stance on a given matter concerning the world and leaves oneself open to the judgment of others. This also necessitates taking responsibility for one's acts and speeches in public. Every time one acts, one becomes visible and shows that one cares about this world and that one takes responsibility for the world. This visibility or appearance is only possible where there is plurality. This plural, ambiguous realm is the realm of appearance, where one can establish oneself as distinct and unique. The public space, in Arendt's understanding (as I have frequently mentioned), is the only space where there can and shall be plurality.

Arendt's reinterpretation of *Dasein* allows her to prove that authentic political subjectivity can only occur in the public realm, shared with other subjects. These relationships are influential in who we are becoming throughout our lives.

The Interconnectedness of Political Subjectivity and Intersubjectivity

How is Arendt able to define an intrinsic relationship between the political subject and intersubjectivity? For Arendt, according to Topolski, "the self can only answer the question 'who am I' by interacting with the other" (Topolski, 2015, p. 18). This points to the intersubjective character of political action. A political subject, for Arendt, can only *become* a political subject through *actively taking part* in intersubjective relationships that take place within the world (Loidolt, 2018). Simultaneously in this process, the person has access to the world perceptions of other people with whom they share the world. Arendt explicitly establishes plurality as "plurality of subjectivities" (Loidolt, 2018, p. 151), which, by definition, includes "plurality of subjectivities that needs to be articulated in worldly activities" (Loidolt, 2018, p. 152). In that sense, plurality, as it is explicated by Arendt, is more than a condition or structure that is necessary for the political realm to be erected; it is a "plurality of first-person perspectives that can be experienced as such in acting and speaking" (Loidolt, 2018, p. 153). As I have repeatedly mentioned, this first-person perspective, the "who" of someone, can only be enacted within the world that we share with others. These interactions are carried out with other people, who share the world, together with oneself. This "who," as Loidolt explains, that

> shows itself—and, at the same time, eludes the fixation of the "what"—is that of acting and speaking: It is intersubjective interaction. The "who" that appears in this interaction is not a representative, nor a reflection of an already full-fledged substantial "inner self"; as an appearance it "'expresses' nothing but itself, that is, it exhibits or displays" (Arendt, 1977, p. 30). Thus, "who one is" only develops in actualization with others. The self is given as self-appearance in the face of others, which, however, does not mean that it is originally individuated by others. Instead of the classical "dative of appearance" that remains in itself, Arendt's brand of subjectivity is turned "inside out" and enacts itself in the world.
>
> (Loidolt, 2018, p. 182)

So, for Arendt, people are "not first in the world and then sometimes appear in it and sometimes do not, but being-in-the-world is essentially entering the world by appearing in it and leaving it by disappearing from it" (Loidolt, 2018, p. 75). According to Loidolt, Arendt shows that "it is precisely public existence that allows one to become oneself, namely in appearing before others and acting together with them" (Loidolt, 2018, p. 31). Thus far, therefore, the way in which Arendt configures her notion of the political subject is as follows: (a) the political subject, the "who," is not a projection of the person who acts in the world; and (b) the "who" is a process that takes shape in interaction with others with whom the subject shares the world. In other words, there is no political subjectivity that is not within an intersubjective relationship.

Here, it becomes clear that this intersubjective relationship is constitutive of the political subject. One thing to highlight here is that if being-in-the-world is always already being-with-others (Loidolt, 2018, p. 75), then it is not too far-fetched to imagine a coconstitutive relationship between intersubjectivity, subjectivity, and the world. Loidolt (2018) quotes from Zahavi (2001, p. 166) in order to highlight this coconstitutional dynamic:

> the three regions "self," "others," and "world" belong together; they reciprocally illuminate one another, and can only be understood in their interconnection. Thus, it doesn't matter which of the three one takes as a starting point, for one will still inevitably be led to the other two: the subjectivity that is related to the world only gains its full relation to itself, and to the world, in relation to the other, i.e. in intersubjectivity; intersubjectivity only exists and develops in the mutual interrelationship between subjects that are related to the world; and the world is only brought to articulation in the relation between subjects.
>
> (Zahavi, 2001, p. 166)

Here, referring to Zahavi, Loidolt points out an important insight in Arendt's political theory. The fact that the regions of the self, others, and world belong together suggests that it is not entirely up to the self to be an authentic existence as disclosedness, as Heidegger portrays. The authentic existence as disclosedness—in other words, the political subject—is a process of coconstitutional dynamic between the self, the world, and the other. In accordance with this insight, Loidolt highlights two features of political intersubjectivity according to Arendt:

> The appearance of the "who" needs and sustains a space of appearance. Its medium of appearance is a web of relationships ... The appearance of the "who" together with others is experienced as an end-in-itself and creates a shared reality.
>
> (Loidolt, 2018, pp. 181–2)

Thus, Arendt positions being-with-others as a preliminary condition for political subjectivity. Loidolt here presents a valuable insight into the content and shape of this being-with-others. For Loidolt, the form of intersubjectivity that one takes is conditioned and accompanied by the specific other (Loidolt, 2018, p. 179). What is significant in Loidolt's reading is that this specific *other* is essential for the acting agent to determine how the *world* (the *in-between*) is going to unfold: as either authentic or inauthentic. Therefore, this specific other is influential in the interpretation of the appearances of the world and how the world is going to be appropriated (Loidolt, 2018, p. 179). In other words, the community that we are engaged in and the world that hosts this community are essential in how we are going to become political subjects. In the next subsection, I show that there are two ways of having intersubjective relationships depending on what kind of others we engage with: others that are concerned with the "what" of someone as opposed to others that are concerned with the "who" of someone.

Authentic and Inauthentic Forms of Intersubjectivity

As mentioned, in order to understand political interaction, there is a need to be able to observe the coconstitutional dynamics between the agents that act, the web of relationships that one acts within, and the world as a collective of cultural and material products at the time of action. The web of relationships that one is within is influential not only in how one establishes oneself as an authentic self but also in how one perceives and approaches the world. According to Arendt,

> The realm where human affairs take place consists of a system of relations, which emerges everywhere where men live together. Since men are not just thrown into the world by accident, but are born by human beings into an already existing human world, the web of relationships of human affairs precedes all singular acts of acting and speaking. Thus, the disclosure of the newcomer through speech and the new beginning that is set by action are like threads that are woven into an already existing pattern. They alter the fabric of the web as they themselves affect all threads of life within the web in a unique way. When these threads are spun to an end, they are recognizable again as patterns and are recountable as stories of life.
>
> (Arendt, 1981b, p. 226, quoted in Loidolt, 2018, pp. 186–7)

Here, Arendt refers to the "web of relationships" that defines, determines, and specifies the possible appearances that can unfold (Loidolt, 2018, p. 187). The "web of relationships," as can be derived from the quotation above, are forms of certain human togetherness. Loidolt calls this human togetherness "we," since they entail multiple agents in intersubjective relationships. This "we" always entails that the acting agent has an intersubjective relationship, either with one other agent or with a group of agents. Loidolt highlights that this "we," just like the distinction between "who" and "what" someone is, can be directed toward revealing the unique "who"s within the "we," thus being directed toward plurality. Alternatively, it can be directed toward being a unified voice with one single mind, thus being limited to the "whatness" of "we." In this way, Loidolt extends Arendt's understanding of the "authentic" or "inauthentic" political subject to "authentic" and "inauthentic" webs of relationships. The important insight to remember here is that, for Loidolt, the form of intersubjectivity that one takes is conditioned and accompanied by the specific other (Loidolt, 2018, p. 179). This influences—even conditions—the possibility of the appearance of an authentic "I."

The question now becomes, according to what criterion is Loidolt able to judge if a web of relationships can be called authentic or not? Here, Loidolt gives a valuable insight:

> For Arendt, an authentic form of "we" does not dissolve the uniqueness of each "I." Quite the contrary, it is the necessary medium of their distinct articulation and appearance. The fully developed and most authentic form of "we" is thus the fully developed and most authentic form of "I" and vice versa. Therefore, the highest possible form of a we-community is that which is dedicated to this task—which Arendt calls "the political"—or, at least, which is aware of its vital importance for a human community ... The "we" for Arendt is neither constituted by a common telos nor by a common intention but by appearance, experience, and actualization.
>
> (Loidolt, 2018, p. 222)

The criterion, then, is to be able to observe in a community a certain directedness either toward plurality or toward uniformity. Below, I unpack this further.

The Inauthentic Web of Relationships

For Arendt, the inauthentic web of relationships can be a very strong form of togetherness, but this group, rather than valuing the distinctness and uniqueness of each of its members, is directed toward diminishing the "uniqueness of the subjectivities involved" (Loidolt, 2018, p. 223). As Arendt says, "rules of uniqueness and exclusiveness are, and always will be, in conflict with the standards of society" (Arendt, 1959, p. 53). As mentioned before, "uniqueness implies living and self-aware beings who are able to express their stance" (Loidolt, 2018, p. 176). The key aspect to focus on in this definition of uniqueness is "being able to express their stance" (Loidolt, 2018, p. 176). Humans in this form of human togetherness—in other words, in an inauthentic web of relationships—are subordinated to "the structure of one organism, apprehending them only in terms of their sameness and not in terms of their respective distinctness" (Loidolt, 2018, p. 223). Therefore, there is no expression of one's own stance, since there is no distinctness or uniqueness. Once the possibility of a person becoming visible and distinct through their deeds and speeches vanishes, that person is either pushed toward loneliness and isolation (and not in a way that Heidegger would wish) or is expected to acquiesce in obeying the norms and rules of the group. Arendt's analysis of totalitarianism and the space of the social stems from this conviction.

According to Arendt, totalitarianism "exploded our traditional categories of political thought (totalitarian domination is unlike all forms of tyranny and despotism we know of) and the standards of our moral judgement" (Arendt & Kohn, 2005a). According to Topolski,

> totalitarianism is unique in that it not only dominates the public sphere; it also rules the private sphere, that is, the family, the house, and other such spaces of privacy and intimacy. It does so by means of organized violence and terror according to ideological laws.
>
> (Topolski, 2015, p.44)

Totalitarianism, thus, led to the collapse of our moral standards. According to Arendt, Eichmann is the personification of the people who went along with the wishes of the leader even when they found those wishes to be unacceptable. Arendt, in her interview with Jürgen Fest, indicates that

> [Eichmann] wanted to go along with the others. He wanted to say "we," and going-along with the rest and wanting-to-say-we like this were quite enough to make the greatest of all crimes possible. The Hitlers, after all, really aren't the ones who are typical in this kind of situation—they'd be powerless without the support of others.
>
> (Berkowitz, 2014, p. 194)

This transformation of regular people into criminals under totalitarian regimes illustrates that a society is capable of changing its ethical standards. These ethical standards, which are based on morality, are unreliable for Arendt, since morality is "a set of mores, of customs and manners, which could be exchanged for another set with no more trouble than it would take to change the table manners of a whole people" (Arendt, 2003, p. 43). Therefore, for Arendt, it is personal responsibility, rather than morality, that should be considered at critical junctures. A society becomes an excellent example of an inauthentic community under a totalitarian regime, since totalitarianism eliminates any possibility of personal responsibility for any act. For Arendt, taking responsibility for one's actions is an expression of human individuality and human distinctness. However, society under totalitarian rule would demand that one conform and behave according to the standards prescribed by society.

In Arendt's analysis, a society under a totalitarian regime is thus portrayed as lacking the capacity to take personal responsibility. Yet not even members of modern society, for Arendt, would be willing to take personal responsibility, since modern society is ruled by bureaucracy. Arendt claims that bureaucracy is the "most tyrannical form of leadership," since it is based on "the rule of nobody" (Arendt, 1998, p. 40). When there is no one who rules, there is no one to take responsibility. In this sense, if totalitarianism is a regime of one-man rule, bureaucracy is a regime of no-man rule. However, both are equally detrimental to human plurality, since they both suppress differing, opposing voices by forcing people to conform to the norms of society.

Aydin (2009) articulates the abovementioned distinction between totalitarianism and bureaucracy as being the difference between "People-as-One" and "People-as-No-One" (Aydin, 2009, p. 813). Aydin argues that totalitarianism was able to provide to the people a reliable form of authority that gives power to the people. This resulted in self-identification of the masses with the perceived authority and having a sense of togetherness that ruled out the internal struggles among people within one nation (Aydin, 2009, p. 811). Totalitarianism in the twentieth century was therefore a clear example of a "People-as-One." As a possible remedy that could prevent the doctrine of "People-as-One" from turning into a "People-as-No-One" doctrine, Aydin proposes Nietzsche's vision of a community (Aydin, 2009, p. 813). According to Aydin, Nietzsche emphasizes a society that is organized and regulated through embodying certain common goals and ideals, where individuals and groups are continuously challenging ideologies since they continuously strive for perfection (Aydin, 2009, p. 814). Instead of identifying themselves with an ideology, Nietzsche imagines individuals striving for personal and shared social goals and ideals. The agreement upon which ideals and goals to pursue depends on a community where human plurality is constitutive of the public realm (Aydin, 2009, p. 814). Since these goals and ideals are "the outcome of continuous interaction between groups and individuals, every goal or ideal will be provisional" (Aydin, 2009, p. 814). There is an obvious similarity between Arendt and Nietzsche here in relation to the location and the form of relationships entailed by human plurality as a remedy to "People-as-One." However, there is a significant difference between the two philosophers with regard to the agent's motivation to engage in an intersubjective relationship. For Nietzsche, we as humans seek power. We may envision a future for ourselves and share the ideals and goals of the community we belong to, but we engage because we want "our" ideals and goals to be shared by many. We discuss and struggle because, ultimately, we want our ideals to be accepted by many. For Arendt, by contrast, power can never belong to a group, let alone a person. Power is the end result when this continuous interaction and discussion reaches a moment of decision to act in concert (Arendt, 1998). Arendt envisions a society that could be defined as "People-as-We," where every individual is treated as an end-in-themself. I would now like to continue elaborating on the Arendtian "People-as-We" that is a remedy to "People-as-One."

The Authentic Web of Relationships

The web of relationships in its authentic form is intrinsically related to visibility, according to Loidolt. This is because

> A real "we" can be sustained only by a space of appearance. Joint forces in labor and work do not need the public space because they are united by something else: a common movement, a common purpose. However, a "we" that emerges from addressing each other directly has no other point of orientation than the appearance of the words and actions of others. This is why actualized plurality needs the visibility of an in-between.
>
> (Loidolt, 2018, p. 225)

This means that the authentic web of relationships requires the constant appearance of others and the subject. Because this form of intersubjectivity is based on speech and action, which are both temporal activities, maintaining this temporality as a constant depends on one more thing: power (Loidolt, 2018, p. 225). In the first chapter, I defined Arendt's notion of power as a potential. It is a potential that is constant, but its realization depends on the acting agents. This realization of power can only occur through a specific form of web of relationships: authentic intersubjectivity, which "is close enough to keep the possibilities of action constantly open" (Arendt, 1981b, p. 253, quoted in Loidolt, 2018, p. 226). This means that when authentic intersubjectivity is realized through acting as authentic agents who address one another in speech and action, the maintenance of this form of intersubjectivity depends on power. However, power can only be generated through this authentic intersubjectivity—this form of "acting together" not as a uniform, single-minded mass but as a collective of acting agents that are in agreement.

One thing Loidolt points out is that in Arendt's formulation, authentic intersubjectivity has both agonal and associational aspects in it (Loidolt, 2018). This means that the group of people that has an authentic intersubjectivity does not always necessarily act like a unified group. Sometimes, people can be engaging in speech that would enable the distinct "who" of someone to appear through that person's words and judgments, and that person can encounter an opposing "who" that contradicts them. This is within the spirit, so to speak, of the political realm, where different world accesses represented by different people can appear. According to Loidolt,

> These "players" struggling with one another are the visible actors on what Arendt calls the "stage of the world," where "spectators," i.e. the public, witness, comment on, and judge them. Here again, we have a loosely united group of spectators, from which the formation of a "stronger we" can emerge, like a community united by kinship in view. Hence, what ties the space of appearance together and keeps it actualized is the form of power that transcends particular interests and is directed at plurality itself.
>
> (Loidolt, 2018, pp. 227–8)

The most important distinction between authentic intersubjectivity and inauthentic intersubjectivity is whether the group is aimed at having one single voice that is representative of all the members within the community or whether, by contrast, the group creates or establishes a space of appearance—a space among them that enables all the actors within the group to be seen and heard by the others in the group. A group that has authentic intersubjectivity will not have a unified goal or aim, and it will not be directed toward taking care of a certain need; instead, the very existence of the group will be an end in itself.

An authentic web of relationships, for Arendt, is a political community. A political community does not have a group mentality, nor does it prescribe any specific way of behaving, rules, or norms for its participants. In a group, one is a member; in a political community, one is a participant. The political community, for Arendt, is characterized, first, by its focus on public rather than private interests, as previously explained. Second, it is characterized by its spatial character instead of a shared identity. What unites people in a political community is the space *in between* them, their participation in this space, and the political institutions that make the space possible. This means that this space always has a material dimension. As the web of relationships, this material dimension is influential in the construction of the "who" of someone—in other words, political subjectivity. I return to this point in the next section. Arendt's model of political community is based on civic friendship, which in turn is based on discourse and mutual respect (Arendt, 1998). Arendt sees solidarity as the basis of this civic friendship, since solidarity for her arises from a shared responsibility for the world that is discovered in dealing with others (Arendt, 1998).

With the disclosure of the world through interaction in a public realm, it becomes clear that we share certain "interests." Interest here means "something which *inter-est*, which lies between people and therefore can relate and bind them together" (Arendt, 1998, p. 182). This leads to a sense of "sheer human togetherness" (Arendt, 1998, p. 180), which creates solidarity in shared projects but leaves plurality untouched. Solidarity strengthens the bonds between citizens in the *polis* because it arises out of an awareness of mutual interdependence and shared responsibility for the common world. It comes into existence out of speech and action in a common political realm. Through a citizen taking political responsibility for the shared world together with others, a sense of belonging to that world arises for them.

Until this point, I have only focused on demonstrating the interconnectedness of political subjectivity and intersubjectivity. In the upcoming section, I argue that material products can determine and influence the very intersubjectivity in which the acting agent engages. In the following section, through an analysis of the existing literature and the interviews I conducted with the participants of the Gezi Park protests in Istanbul, I focus on the role of technology in these intersubjective relationships.

The Polis-Making of Gezi Park

During the Gezi Park protests, the park, which was originally created for leisure activities, assumed various roles in line with its meaning for the Turkish people. It became a home for those who erected tents and slept on benches; it became a place to eat for some, since there was a kitchen that produced fresh food; it became a place of medical care and attention. The park provided all of the basic services that a state would provide: healthcare, shelter, food, and in some cases even education. In the subsection below, through an examination of existing literature and the scripts from the interviews I carried out, I show how the park was appropriated by the protesters and transformed into a *polis*, a space of appearance in Arendt's use of the term.

The Material Organization of Gezi Commune

After the first day of the protests, the protesters formed what was called the Gezi Commune in order to stop the police from entering the park again. The Gezi Commune occupied and governed Gezi Park between June 1 and June 15. The material setup of the commune was as follows: there was an empty space in the middle of the park, left intentionally unoccupied; during the protests, this was where people gathered and debated. Several areas functioned as medical help centers where doctors treated those wounded as a result of police brutality. A dedicated kitchen established by UNICEF provided food to the protesters twenty-four hours per day. Multiple food courts operated across the site, supplied by people who attended the protests or supported them by bringing food. Pharmacies were available for all kinds of medical needs. Coordination points helped people decide where to walk to, determine where the police were waiting, and contact media willing to show the protests in their news. A library offered books to read. In some locations, one could also find clothes that had been ripped off or discarded as too bloodied by wounds (Yeşil Gazete, 2013). This settlement was created in two to three days in the park during the protests, and the literature on the Gezi Park protests agrees that it was created spontaneously during this period.

After the borders had been set within which the protest was going to be carried out, services that were necessary for taking care of basic needs were spontaneously established. An organization that consisted of volunteers within the park, called "Gezi Parkı platformu,"[6] was carrying out the services. An individual could enter the park from any entrance and could immediately take care of their needs. In the food courts, one respondent said, "You could find anything that you would find in a small Cuban supermarket, like essential goods." The food services were either provided from small counters, or people visiting the park would bring food with them to give away to protesters.

As well as food being provided for the people in the park, infirmaries were also organized where volunteer doctors and medical students worked every day. The spontaneous, voluntary organization of the protesters was making the park functional in terms of taking care of needs. After the needs were taken care of, there was room for politics to emerge. One respondent described a regular day in the park:

> all day you would sometimes read books or magazines; or sometimes there were activities related to art; there were small theater shows that people were putting out among themselves; there was no need for a stage to be established; on every corner of the park people were gathering together and doing things among themselves. … Generally, it was very orderly, and everything was very clean. For example: people were collecting the garbage in front of their *homes* and taking care of their surroundings; they were asking if you have eaten, if you didn't have extra food; if you are cold, have an extra blanket. … With things like this they created a sense of home in the park, a sense of neighborhood.

When asked what she meant by homes, the respondent changed the word into tents. In a follow-up question, when asked why she confused a tent with a home, she responded:

> people were feeling like that place is their home because they were feeling very secure there. I mean, yes, sometimes we don't feel at home living in this country, especially in these latter days. You are living, walking the streets with the feeling a bomb can explode, and you can die, and this is something that a person should not be feeling at home! A person should not be scared in their own home, should feel secure. Everybody was feeling secure in the park, for example the phone, no one lost anything there, but even if I lost my phone, I wouldn't say either a homeless kid or the person in the nearby tent stole it; I would think I either dropped it in the bathroom or in the line for food. The park created that sense of security, that sense of being at home.

This quotation is very important since it designates two interconnected phenomena related to how politics was exercised in the park: on the one hand, while being a public space in itself, the organization within the park allowed protesters to claim their right to privacy. In addition, they could withdraw themselves from participating in the public life of the park by going into their tents. On the other hand, the park allowed the participants to experience a sense of togetherness by sharing what they had with one another in every manner, from needs-based goods and services to thoughts, art, and music. The respondents said that there was a stage that was placed in the middle of the park, a little to the left of the pool. In the area where the pool is, the circular space within the rectangular area was appropriated as a space for public speaking of words and public performance of deeds. People were not confined to executing *praxis* in that area (they were already doing so in their encounters with one another), but this was a space designated specifically for that. A concrete, open space enabled itself to be appropriated into a space of appearance, and it thus turned into a public space.

It is almost intuitive to think that social or political interactions need an appropriate place to occur. However, the factors that affect these social or political interactions, such as the design components of the space itself that influence the possibility for these interactions to occur, do not receive enough attention.

Urban scholars argue that there is a direct relationship between cultural diversity and public parks (Low et al., 2009). Low et al. argue that the more its urban parks, beaches, and heritage sites remain public, the more a society will be able to integrate diverse communities and enable social tolerance. This is because, Low et al. claim, these spaces allow people to have a place in which "interpersonal and intergroup cooperation and conflict can be worked out in a safe public forum" (Low et al., 2009, p. 3). Low et al. define these spaces as "places where a wide variety of people of different gender, class, culture, nationality, and ethnicity intermingle peacefully" (Low et al., 2009, p. 4). However, these approaches fall short in explaining the role of public spaces in the creation of certain publics. Gezi Park and the Gezi Park protests prove to be the perfect case to explore the role of technology in the form of the park itself in order to unravel how the park enabled a specific public to occur.

According to Karayalı and Yaka, Gezi Park, through forming the Gezi Commune, was an active agent in the protests:

> The primary infrastructure of the Gezi Uprising—the material setup that *facilitated* and *enabled* its constitutive modes of action and interaction—was the park itself. As with any modern park, its paths encourage pedestrians to drift, to wander in circles and ultimately to decelerate. In contrast to Taksim Square, right next to the park, the space isn't suitable for marches and the choreography of unified masses. Rather, the fabric of the park facilitated a kind of federalist mode of assembly, enabling encounters between different sections, groups and identities who could thereby relate positively to each other, in the first step towards recomposition.
>
> (Karakayali & Yaka, 2014, p. 124)

Gezi Park, then, was not only a political mediator (because it is a materially stabilized realization of the memories of what happened in that space before), but it also facilitated and enabled the conditions of possibility for plurality, equality, and freedom to exist as a worldly reality. Below, I explore the community that was created during the protests. I argue that this community is the example of the authentic "we" that I defined at the beginning of this chapter.

The Gezi Commune

As I have mentioned several times already, the Gezi Park protesters did not have a common ideology; they were not one homogeneous group belonging to the same political faction, ideology, or party. The protesters were composed of diverse groups of people with different backgrounds, levels of education, and ages. Eken, observing that there was an obvious heterogeneity of the people within the park, defines some of the participants as "communists, anticapitalist Muslims, LGBT groups, anarchists, feminists, nationalists, Kurds, unaffiliated women and youngsters" (Eken, 2014, p. 434). One of Eken's most important insights is that, while the park housed these different identities and groups of people, *the ambience* within the park proved that holding on to these identities and groups was "utterly inadequate during the commune" (Eken, 2014, p. 434). The people within the park, according to Eken, were not very concerned with what one is or what one reveals; it was more important to observe who one becomes (Eken, 2014, p. 434). The people who were attracted to Gezi Park and wanted the park to remain as it was had different backgrounds in life, different experiences, and different wishes for their future. Yet Gezi Park engulfed all these backgrounds, experiences, and wishes into something very common to all of these people: freedom. Because the focus of the people within the park was not on who is what, but rather on sharing experiences in the manner of interaction, this interaction allowed people to encounter the other, while they themselves were becoming others too. My meaning here can be clarified by quoting what one of my respondents said:

> for me, until Gezi, the hijab was a political symbol; maybe this is because of my own conservatism or standardization. But after Gezi, the hijab was not a political symbol for me anymore because I saw women with hijabs at Gezi too.

One might have strong ideas about a subject, like the hijab symbolizing the conservative Islamic lifestyle that opposes secularism, and it had indeed been utilized as such by many political parties in the past in Turkey. However, not all women who wear the hijab are wearing it to symbolize a choice as such. As already mentioned, the Turkish political environment was so polarized and fragmented before Gezi that people did not have the room to test their convictions, beliefs, and ideas. Gezi opened that space of encountering the other while also encountering oneself. As Eken argues, the solidarity and commonality in the park during the protest were created out of this potential and possibility of one *becoming*, transforming, and changing in relation to the interactions in the park (Eken, 2014, p. 434). Eken continues:

> This is the logic of becoming: becoming is based on the seeds of transformation planted in an encounter in which each party moves on to something else that he or she could have never formulated before or outside the encounter. Thus an identity can never become; rather, one becomes only through a minoritarian condition as what resists identity as a majoritarian norm.
>
> (Eken, 2014, p. 434)

In the previous chapter, I discussed how Gezi Park mediated the common sense of its time of creation—being a citizen rather than a subject of the Sultan. This common sense of the new Republic of Turkey at the time still has a strong place in the minds of Turkish people. Whatever political affiliations they might have, they know that they have the right to freedom and do not have to bow to the Sultan anymore; they are free citizens. This also becomes clear from the writing of Ahmet Insel, the famous Turkish academic and columnist, during the Gezi Park protests.

In his column in the newspaper Radikal, on June 4, 2013, Ahmet Insel (2013) argues that the Gezi movement is not similar to the movements within the Arab Spring. For Insel, the Gezi movement is a "revolt of dignity," since the people who are attending the protests cannot be summarized as being against certain policies of the government (2013). According to Insel, this movement is

> a reaction against a person who boasts about the absolute power of the government, who perceives opposition and criticism with the codes of a seams braggart and who grew up in the culture of submissiveness. This is a revolt of people who cannot tolerate this aggressive and imperious style of politics and debased attitude, and who believe that their dignity of citizenship is damaged.
>
> (Insel, 2013)

Insel's analysis is also in line with what one of my respondents observed while being in the park. According to my respondent,

> Ahmet Insel called this a revolt of dignity; I believe he and I are making a similar analysis, with simple words. It is like saying, "Don't butt in to my life, don't butt in about the length of my skirt, don't butt in about who am I going to live with, make love with, If I live in a co-ed or I don't, don't butt in to these things." … If we separate the first term he served, Tayyip Erdoğan and his governments interfered in every aspect of our lives for 12 years, and they still continue to do so. And in the Gezi movement, people said, "Enough, don't interfere." He talks every day, every day he says something about our lives. If you look at the crowd at Gezi, everyone has a different … their own resistance. For years they didn't have the room to resist, and they caught the possibility at Gezi. It is almost childish, but one of the mottos of the park in those days was, "That's enough, shut up."

This, I believe, is a good summary of the sentiment among my respondents when asked why they were at Gezi Park. It was a claim to freedom and a claim to dignity: being able to choose what one wishes for one's own life rather than being told what one's life should be. Even though people were not always in agreement with one another, the political climate of the park was such that an individual respected the other who did not think or live like them. One respondent said, "I remember the first day we took Gezi Park, I remember how kind and understanding people were, the priority was human, I cannot say anything else, it was focused on humanity, nobody cared about politics."[7]

When I asked my respondents about how politics was experienced in the park, none of them talked about power relations between factions within the park, and none of them talked about reaching a goal or about organization toward reaching a goal. Their main concerns were for the presence and discussion of interaction, self-expression, and being able to live together where there are different views on how to live. Based on their participation in the Gezi Park protests, Bilal and Nahrwold (2013) reflect on their experience:

> People experienced a new solidarity in the Gezi Commune. It was an alternative community, with a new usage of public space. The protesters reclaimed their space and experienced an urban utopia. In the Gezi Commune, there was no state violence because police were not allowed entry, no cash flow because everything was free, no authoritarian politics but instead directly democratic forums. An ideal public space was created, which empowered different perspectives and made possible something unique: different people contributed with their individual abilities and made themselves agents of history.
>
> (Bilal & Nahrwold, 2013, p. 91)

This common sentiment of a claim to freedom, according to one of my respondents, created a sense of solidarity among the people in the park. According to this respondent, the park was

> a space that we governed. And nobody wanted people to drink alcohol and go nuts in there; freedom didn't mean going crazy. … I believe there is something problematic in the thing that we call the state, but during the Gezi protests small organizational states appeared. I mean they were trying to regulate and organize things, and those were causing factions among people, but … people were taking care of one another there. There was solidarity among people. The motivation to control yourself and act in a more civilized way became more apparent—I really liked that—and this sense of solidarity is a very nice thing. If you asked for help, you would immediately get it. There, your act is not necessarily individual, but it is something that melted within something bigger, social movement is this anyways, but that sense of brotherhood and acting together really made an impression on me.

The respondents described some common experiences of feeling free—feeling respected and accepted for who they were—but their experiences varied. Some respondents said they learned a lot about opposing points of view, which helped them critically assess their own stances. Others said that they became more prejudiced against Erdoğan and his supporters.

These commonalities and differences were present in almost all my respondents. They had different concerns, and they had different perspectives on how to be and how to live. However, unlike at any other time in Turkey, this time, they were able to express what they thought about this experience, show themselves as unique beings in relation to this experience, and be challenged by others in relation to this experience through interaction. One of the mottos of the protests was inspired by the graffiti that could be seen on walls all over Istanbul saying "Kahrolsun bağzı şeyler!" which would be translated as "Down with something or other!" (Seloni & Sarfati, 2017, p. 16). According to Seloni and Sarfati, this and many other uses of political graffiti point to how the protesters "tried to symbolically and linguistically blur the ideological distinctions between polarized groups, such as Kurds and Turkish nationalists or Islamists and secularists" (Seloni & Sarfati, 2017, p. 25). According to Arda:

> other ways of interaction were promoted among the protesters in the Gezi Commune. This not only means that alternative behaviors were considered "valuable" but also that the terms of public appearance changed when the system of delegation lost its validity. Every protester was involved in the protests

> through their own personal commitment, without the label of identity, class, or any group affiliation, since these notions of entity did not apply in the sphere of the Gezi Commune. This indicates the emergence of a new "common" and the production of social relations diverging from the existent system.
>
> (Arda, 2015, p. 16)

By blurring these ideological lines, the protesters opened the room for discussion, allowing everyone to voice their own grievances. When I asked one of my respondents what the "something" in the graffiti referred to, he talked first of all about the authorities' plans to cut down trees in the park. But the slogan meant more than this to him, as he explained:

> According to me, that something was not the tree, it was not the park. That something there was intervention into lifestyle and generally opposing the AKP. I mean I don't believe, of course, there is some sense of sensitivity towards the tree and the environment; what they did there is a serious massacre; it is not one tree there. But the problem is not the tree. Like the AKP says, "Look, the problem is not the tree; they want to overthrow us." I agree with this; they read this differently, according to their own agenda, but really the tree was not the only problem there. Yes, the starting point was the tree, it was nature. But we need to look at the conjuncture at the same time: there is a third bridge[8] and the name of the bridge is Yavuz Sultan Selim.[9] There is hopelessness everywhere; the system is not working; democratic institutions are not working; this government cannot be overthrown or weakened, or it does not act according to the rules of democracy while being in power, it creates its own one-party state. The people in the park said, "Stop!" to this. As a general umbrella, it was being against the AKP.

This respondent claims that most of the people were in the park because they were against the Justice and Development Party (AKP). It comes across implicitly from my interviews that the reason for people to oppose the AKP was the intervention of the ruling government into citizens' daily lives. Resisting these interventions also went hand-in-hand with identity claims, as some of my respondents perceived themselves as secular, liberal people who prioritized personal freedom in the sense of being able to do what one wants, as long as one does not harm others. While some of my respondents perceived an attack on their identity to come in the form of a ban on alcohol, others saw Erdoğan's comments on women's clothing and his call to have at least three children as an attack. The more people were able to express these grievances, the better they were able to understand one another. One respondent said,

> Gezi says, "Down with something or other." This is a message that also has a negative content outside of ideology. It said, "Don't do it, don't touch my park." That was the first thing the protests said: "I won't allow it, you are doing wrong and I am doing right." … The AKP was able to beget victimhood and claim a moral superiority until the Gezi protests, but there they lost it because they attacked innocent people. Gezi said, "I won't allow it"; that is the only thing Gezi said. It didn't say anything other than that or even try to say anything other than that.

The sentiment that this respondent expresses about the protesters, namely "I won't allow it," is a very concise summary of the matters of concern of the people I interviewed. Some were deeply concerned about what they saw as the neoliberal urban transformation of Istanbul, while others were concerned that they would lose their individual freedoms. Some were concerned about keeping the park intact and making sure others remembered the past the way that they did, while others were concerned about the autocratic tendencies of the government.

The aim of demolishing Gezi Park was not merely for commercial or cosmetic reasons. The government wanted to take down the carefully spaced trees that allowed people to see others while walking in the park. At the end of the walking route, the government wanted to build on the circular area where park benches are located, where people could encounter one another and possibly interact if they wished to do so. The intervention of the government in attempting to demolish this public park was an intervention in the political landscape of Turkey, since through its design and its place in the common sense of the Turkish people, this park could turn into a space of appearance at any time. That was the danger the government wanted to eliminate: the possibility of humans acting—and acting together—since it is that form of action and power that can begin something new, that can start revolutions, and that can overthrow governments. There was a dynamic relationship between the park, the commune, and the individual acting agents within the park that all played a role in the formation and cultivation of a new public.

Based on the analyses above, I argue that the organization of the protesters within the park during the protests portrays a specific interpretation of sharing a common space. This specific interpretation of sharing a common space is fostered by spaces that are multistable. In the section below, I demonstrate that the formation of a new political community in Gezi Park was actually an end result of technological mediation of intersubjectivity.

Technological Mediation of Intersubjectivity

In this section, following the theoretical and empirical insights already discussed, I demonstrate technopolitical mediation of technologies as a two-step process. I show that Arendt's definition of public space (which requires plurality of perspectives and a stable material space) is inextricably linked to the concept of multistability. I argue that spaces that are multistable (and maintain their multistability) enable the formation of an authentic political community. Providing a new interpretation of Kudina's hermeneutic lemniscate, I show that the material space in which people meet mediates both the way in which a community is formed and the way in which one becomes a part of such a community.

Gezi Park and Multistability

As has been frequently mentioned, people are drawn to certain physical public spaces because these places have *meaning* for them, they remember the events that happened in those places, and those events mean something for them as individuals and as a community. As mentioned before, Arendt positions common sense as the basis of our common cultural context, but this common cultural context is also informed by physical, material products like public parks or buildings. Let me return to the quote where Arendt herself on one occasion acknowledged this:

> If … we understand the political in the sense of the polis, its end or raison d'être would be to establish and keep in existence a space where freedom as virtuosity can appear … this is the realm where freedom is a worldly reality, tangible in words which can be heard, in deeds which can be seen, and in events which are talked about, remembered and turned into stories before they are finally incorporated into the great storybook of history. Whatever occurs in this space of appearances is political by definition, even when it is not a direct product of action.
>
> (Arendt, 2006a, p. 153)

In this quotation from Arendt, it can be seen that she is portraying a definition of a public space: a space that hosts freedom and where the manifestation of this freedom turns into a story that is embedded in the space where it occurred. Thus, the presence of multistability—both multiplicity and stability—is essential to be able to call a space a public space. However, what should be remembered here is that meaning making for Arendt, as has been seen, is not an individual that enables meaning making. The concept of multistability, as it has been explored in postphenomenology until now, does not take into account this pluralistic and communal act of meaning generation. In order to explain my point further, I need to revisit the concept of multistability.

Multistability Revisited

In the previous chapter, I touched upon the notion of multistability, since it is one of the building blocks of postphenomenolgy.[10] Don Ihde was the first philosopher who conceptualized the way in which we can think of a technology that is "designed for a specific purpose but is at the same time open to alternate uses and yet at the same time again still somehow remains limited in the ways it can possibly be used" (Rosenberger, 2017, p. 10). The concept of multistability explains the capacity of technologies to be used for different purposes than those for which they were designed. Ihde argues that "no technology is one thing, nor is it incapable of belonging to multiple contexts" (Ihde, 1999, p. 47).

However, this does not mean that any technology, device, or artifact can be used for any purpose desired and can be meaningful in any way possible (Rosenberger, 2017, p. 14). It means that the design of the technology is influential in the possible ways in which that technology can be used. The design of an artifact shapes the uses and meanings that are possible in relation to the use of that artifact. This means that, in relation to the design of the technology, "only some relationships between a user and a technology prove to be 'stable' within the experience of users, and a user's relationship with a technology is open to multiple 'stabilities'" (Rosenberger, 2017, p. 15). Multistability shows that technologies always carry within themselves the potential to mediate human experience in various ways (Rosenberger, 2017, p. 16). Whatever the intended use of the technology, there is always more than one "path for a given technology to transform a user's ability to act on, perceive, and understand the world" (Rosenberger, 2017, p. 16). The material structure of the artifact, though, presents limits to the potential interpretations, acts, perceptions, and understandings of the users (Rosenberger, 2017).

Thus, to sum up the two main points that multistability highlights, it can be said that (a) it is always possible to have multiple relations and uses of a technology, and (b) this possibility is also limited to the material configuration of the technology or artifact in question (Rosenberger, 2014). According to Rosenberger (2017), "a device like a park or subway bench is simply an inert object" (p. 5), meaning that these objects do not have a mind or intentions. However, Rosenberger continues, these objects are still capable of being part of "large-scale collective ends" (p. 5). Here, Rosenberger highlights the necessity of understanding technologies as things that are open to certain uses while being closed to others (p. 5). This allows technologies to be "amenable to concrete alteration by different social forces advancing different political objectives" (Rosenberger, 2017, p. 5). Taking as an example Isomu Noguchi's 1933 architectural creation Play Mountain, Wittingslow argues that artifacts can actively and intentionally be designed as multistable (Wittingslow, 2020). This allows the possibility of a park being appropriated as a public space, playground, strange landscape, and sculptural installation, according to Wittingslow (2020). But then, how do we know, in our collective consciousness, that a space has been appropriated as such? Does other space have a place in our collective interpretative framework? Rosenberger connects this issue to visibility, as I explore below.

Dominant and Alternative Stabilities

Rosenberger argues that the stability that is the second half of the term "multistability" should be analyzed further, since it can be a key to understanding how certain objects come to be used in the same way by generations of people. This type of stability is called the "dominant stability" of an object (Rosenberger, 2017, p. 17). Dominant stability is the main purpose for which a technology is created and also the purpose for which it is used by most people (Rosenberger, 2017, p. 17). Even though technologies can assume multiple stabilities, dominant stability refers to "the context of the most frequent use of the technology in question" (Rosenberger, 2017, pp. 17–18). To explain multistability and dominant stability at work, Rosenberger presents his famous example of the public benches. A regular bench is a simple design; as described by Rosenberger, "it is designed for the simple purpose of providing a place to sit. It accomplishes this by presenting a horizontal surface of appropriate size and height" (Rosenberger, 2017, p. 8).

This type of public bench, according to Rosenberger, is common to parks, train and subway platforms, and bus stops (Rosenberger, 2017). Like most technologies, public benches have a dominant stability, which is the purpose for which they were designed (Rosenberger, 2017). The dominant stability of the public bench is that it is a place to sit, and the bench mediates its users' experience by turning a regular space within the world into a place to sit (Rosenberger, 2017). The alternative stability of the public bench can be a bike rack or a support for leaning against for joggers to perform stretching exercises (Rosenberger, 2017). Rosenberger adds that there is another well-known use of public benches: They are also used by homeless people to lie down on (Rosenberger, 2017). For example, a bench that is placed under a bus stop with a small roof can be a space of shelter to the homeless in three ways (Rosenberger, 2017). It is a place to hide from wind, rain, and sun; it provides the relative safety of a crowded and well-lit area; and it provides an occasion to circumvent anti-loitering laws (Rosenberger, 2017).

According to Rosenberger, this alternative use of public benches is disappearing due to a design choice: adding a vertical seat divider, armrest, or handle to the long plank where people are supposed to sit (Rosenberger, 2017). Rosenberger suggests that this design choice is working hand in hand with the anti-loitering laws (Rosenberger, 2017).

From Public Benches to Public Spaces

For Rosenberger, the example of the public bench, with its dominant and alternative stabilities, is an excellent example that can be expanded to public spaces in general (Rosenberger, 2017). There are various dominant stabilities of public parks: They are places of outdoor recreation and relaxation, which allow people to exercise and gather together (Rosenberger, 2017). They are spaces for events, and they are vital green spaces within cities that provide playgrounds for children (Rosenberger, 2017). These dominant stabilities of public parks are accompanied by alternative stabilities. They can be places of shelter and sleep for the homeless, for example (Rosenberger, 2017).

Through the example of the benches, Rosenberger aims to highlight that technological design can serve a certain political agenda (Rosenberger, 2017). As with the example of Gezi Park I have provided in Chapter 3, Rosenberger argues that specific design choices of technologies can invite or inhibit specific actions in relation to them. It can be readily understood how Gezi Park was an example of inviting a certain action (being able to enjoy a free, public life as a citizen), and Rosenberger's benches are an example of how a design choice can inhibit a certain action (sleeping on a bench). According to Rosenberger, any technology, whatever its design, is open to multiple uses and meanings

(Rosenberger, 2017). However, social agendas of city officials, governments, or even society in general have an influence on the multistability of technologies, and this influence takes the form of certain modifications (Rosenberger, 2017). Rosenberger outlines these as restrictionary, unrestrictionary, and expansionary modifications (Rosenberger, 2017, p. 36). In restrictionary modifications, the undesired alternative stability of a technology is closed off. According to Rosenberger,

> A powerful social group may prefer to restrict the device in a way that renders it possible to use only through the group's own preferred stability. Through material inscriptions, the technology can be modified so that it is difficult or impossible to use in terms of the undesired stability.
>
> (Rosenberger, 2017, p. 36)

The modifications to the public benches are an example of restrictionary modifications for Rosenberger. There are two subcategories within restrictive modifications: Prohibitory modifications aim at making it impossible to use a device in relation to a particular use, and dissuasionary modifications only aim at discouraging that usage (Rosenberger, 2017, p. 36).

Unrestrictionary modifications occur after a restrictive modification has been made and a certain stability of a technology has been closed off (Rosenberger, 2017). In this case, "a rival network may work to 'reopen' that stability" (Rosenberger, 2017, p. 37). Using "counterinscriptive methods," the opposing group can lift the restrictions and make the closed stability a possibility again (Rosenberger, 2017, p. 37). Rosenberger's third category is expansionary modifications, where technology is modified to expand the multiplicity of stabilities that it has (Rosenberger, 2017, p. 37). Expansionary modifications give room for different interpretations and different possibilities of use (Rosenberger, 2017, p. 37). In this category, according to Rosenberger, a distinction can be made between a facilitationary modification, which simply allows a new stability, and a persuasionary modification that will "actively encourage usage in terms of that new stability" (Rosenberger, 2017, p. 38).

Through the example of the wall-mounted subway seats in Montreal (where the designer intended the low seats to be occupied by children, but the design worked as an anti-sleep public bench), Rosenberger highlights that the effect of the design of a technology can be separate from the intent of the designer (Rosenberger, 2017, pp. 62–3). This shows that in order to understand the impact of the design of a technology, one needs to look at the societal structures, history, and culture where that technology is embedded. As explained through the example of anti-sleep public benches, technologies can aid and implement certain agendas of certain social groups, and they can make another group invisible within public life (Rosenberger, 2017, p. 68). Here, Rosenberger argues that if, for example, the people of a city do not encounter homeless people in their daily life, then the problem of homelessness becomes invisible.

According to Rosenberger, the success of these restrictionary modifications lies in their capacity to make problems such as homelessness invisible. Once something is invisible, this "has the potential to lead the larger community to grow unaware of the problem or even to mistakenly assume that the problem is less severe than it actually is" (Rosenberger, 2017, p. 68). Once people no longer encounter the problem in their everyday life, it becomes easier for them to be oblivious of it and to never even think about it (Rosenberger, 2017, p. 69). According to Rosenberger, technologies should be understood "as essentially wrapped up within our society's larger politics, including economic systems, law enforcement procedures, democratic and undemocratic representational schemes, penal methods, and racial and sexual power dynamics" (Rosenberger, 2017, p. 75).

Technological Mediation of Intersubjectivity in Gezi Park

Until now, I have delineated Rosenberger's approach to multistability. In particular, Rosenberger's unpacking of dominant and alternative stabilities of technology is a valuable addition to the concept of multistability itself. It is obvious that these alternative stabilities are end results of the appropriation of technologies. However, Rosenberger's argument has the same shortcoming of describing appropriation (meaning making) as an individual activity. It is true that meaning making can be individual. As Rosenberger convincingly shows, by appropriating a public bench as a place to sleep and as a space of shelter, a homeless person is able to create alternative stabilities of a simple technology. However, these alternative stabilities do not have a place in our collective consciousness because these alternative stabilities are *made* invisible, and therefore, the meaning attached to these public benches is erased. What Rosenberger misses is that meaning creation, especially when it concerns *making an appropriation visible and stable*, is a collective act.

If Rosenberger's insights are applied to what happened in Gezi Park, it can be claimed that Gezi Park assumed a different identity—a different role in people's lives—during the protests. As mentioned, the dominant stabilities of the park are its uses as a space of relaxation, as a green area, and for a variety of leisure activities for the people of Istanbul. Taksim Square is also a space that has been appropriated for different purposes and by different groups of people over time. In short, it is the gathering spot for different interest groups to raise their voices, for whatever reasons they may have (Gül et al., 2014). Therefore, the park has a strong affiliation with what Taksim Square means for people in Turkey: freedom, equality, and diversity. However, as has been seen, because the park was maintained as a public space where freedom, equality, and diversity are the driving values that engulf the community, the park during the protests assumed an alternative stability: a *polis*.

Based on the organization of the protesters within the park, Gezi Park during the protests constituted a specific interpretation of sharing a common space. I believe this interpretation illustrates a close relationship between the authenticity of a community (and eventually of the individual) and the concept of multistability. A community could be considered authentic when it creates a new meaning through the appropriation of an artifact while allowing differing interpretations to coexist and not simply adapting the dominant stability of the artifact in question. The inauthentic community would be one that embraces the dominant stability of an artifact while suppressing alternative interpretations and appropriations of the artifact in question. The protesters' interpretation of sharing a common space was itself the process of technological mediation of intersubjectivity: The interplay between humans and the material design of the park opened new realms of action within the world through fostering engagement between diverse subjectivities. However, the fact that this park was able to be interpreted as a *polis* was dependent on its multistable character.

As Wittingslow (2020) argues, not all public spaces are multistable. For example, one could imagine an urban ethos that is both deterministic and suffocating; one that, despite the existence of public space, does not allow for creative or playful reinterpretations of function and purpose. The objects inspired by this ethos might well hold its inhabitants in the world, but this ethos would also make meaningful political action impossible and might actively jeopardize the appearance of political action as Arendt intended the term (Wittingslow, 2020). Nor is it the case that all multistable spaces are public spaces. As previously noted, it is political interaction that turns a space into a public space. Not all human communities gather in a space for political interaction. In that sense, to be able to observe technological mediation of intersubjectivity, it is first necessary to be able to observe technological mediation of common sense. Even if the mediation of common sense occurs for someone who engages with a technology, how the interplay between the subject and the intersubjective community will play out is to a certain extent dependent on the multistable design of the technology in question. In addition, the existence of a political community that appropriates a space as a public space is also dependent on the context of the mediation of common sense.

In that sense, I argue that in order to grasp the political significance of technology, it is necessary to go through a two-step analysis of the technopolitical mediation: first, the mediation of common sense, and second, the mediation of intersubjectivity. In the following section, I illustrate the two-step model of technopolitical mediation through an elaboration of the material space of Gezi Park and Kudina's model of the hermeneutic lemniscate.

Technopolitical Mediation

In the previous chapter, I explained technological mediation of common sense through the case study of the Gezi Park protests. With this case study, I demonstrated the move from the upper left curve toward the lower right curve of the lemniscate model (Figure 3).

In proceeding from the upper left curve to the lower right curve (as portrayed with gray arrows), it can be seen that human access to common sense and the common world was mediated through technology, in this case the physical space of Gezi Park itself and social media platforms like Facebook and Twitter. Gezi Park strengthened the common aspect of the world and brought these people together for its preservation, not only because it was in danger but also because

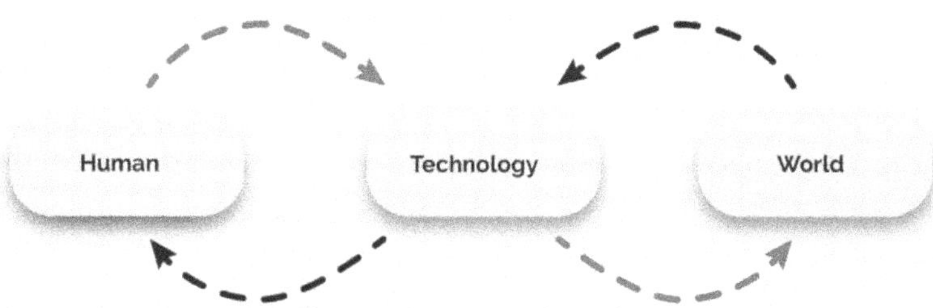

Figure 3 The Hermeneutic Lemniscate (the first move).

of its importance to the people of Turkey. Social media reinforced this existing reality and enabled access to the world through the spread of information. Gezi Park conveyed common sense, as it is the material form of Turkey's cultural memory: The Turks knew what Gezi Park stood for because of their common sense. According to Arendt, a space only becomes a public space through human appropriation. Thus, this human appropriation is not in vain—even in the case of Gezi Park, where the participants were eventually dispersed by the authorities—because it has achieved this spatial transformation. Only when a person appropriates a space through their *praxis* does this space become a space of appearance.

Up to this point, I have given an overview of what appealed to my respondents and made them act or understand. Until this point, Kudina's lemniscate is helpful, flawlessly illustrating the first step of technopolitical mediation. The problem, however, occurs on the move from the upper right curve toward the lower left curve. For Kudina, at the point where the world is, the human being is a singular person. It is true that this singular person has access to their personal history and the history of the community that they belong to, but nevertheless the person in this scheme is alone. The movement from the world to the person through the technology in this lemniscate is concerned with a specific meaning making of one person once engaged with a technology in order to have access to a phenomenon. Kudina (2019) argues:

> through the effective histories of people, the initial meaning of a technology remains in flux. As soon as a preliminary meaning is established, it becomes the background for further interpretation, triggered by the new information about the device, new or possible contexts of application and the experience generated in using the technology. The process of appropriation is never finished. It is the effective history of a person, together with the productive context and the technological mediations, that ensure multiple possibilities for interpreting the same technology, and consequently, devising several courses of action.
>
> (p. 105)

As she clearly shows, this moral change is strongly related to the cultural and moral history of the person in question. The problem is that the meaning making—even the act of establishing the preliminary meaning, let alone generating a new meaning through appropriation in a new technology—is not a singular activity. In the previous chapters, I showed that access to common sense, in other words, access to the interpretative framework of a community, is already an act that one does not carry out alone. One is always embedded in the community that one belongs to. Kudina presents techno-moral change as an inward activity in which a singular person engages with the world through a technology. This person certainly has access to their individual history, but that is all. This new meaning, or the new stability that is reached with the technology, for Kudina, does not even need to be seen or heard by others; it is already within the world.

I argue that Kudina's model of the hermeneutic lemniscate needs to be expanded in the direction of visibility and intersubjectivity, and it needs to be construed as a two-step model to clearly signify the technopolitical mediation of technology, for two reasons. First, as Rosenberger (2017) shows in his analysis of the public benches, the appropriation of a technology needs to be visible in order for that appropriation to have its place among multiple stabilities that the technology can possess. This visibility refers to the collective aspect of public life. Kudina's approach fails to acknowledge the collective aspect of meaning making and of meaning stabilization. It is true that once a meaning is stabilized, this does not mean that the technology cannot be appropriated. However, for the world to incorporate the new meaning that was generated concerning the technology in question, that meaning has to be seen or heard: It has to be an outward activity. And any outward activity—any public act—is necessarily influenced by others. Second, this outward activity can only be carried out by people who have already gone through the technopolitical mediation of common sense. It is appropriate here to return to this argument with reference to the hermeneutic lemniscate.

The discussion now concerns the move from the upper right curve toward the lower left curve (as portrayed with light gray). After entering Gezi Park and reaching the protests, a person would have seen numerous groups of others that they had not seen before: a crowd that was attracted to the same phenomenon in the same space as this person was. We already know that the individual who

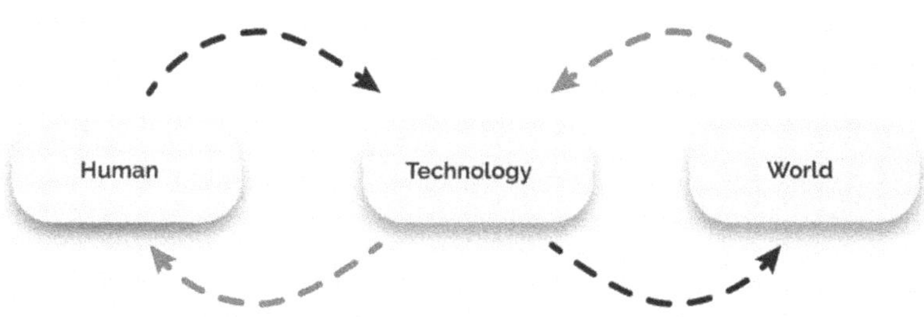

Figure 4 The Hermeneutic Lemniscate (the second move).

was attracted to the park had in their mind one of the amplified meanings of the space, which is freedom.

With this in mind, and because of what the park allows the individual to do, the individual was able to transform a common leisure space into a public space. Because Gezi Park and its surrounding area mean freedom, one who is attracted to that place also acts in accordance with the notion of freedom and shows respect for freedom of expression. This informs all the actions one carries out while being in the park. In other words, the attitude of an individual toward others—respecting the others' rights to freedom and equality—is, to an extent, an end result of that individual's engagement with the park and what the park means. I explained this engagement as the first step of the technopolitical mediation, which is the mediation of the common sense.

The people in the park were moved toward interaction rather than dispute, listening rather than shouting, and understanding rather than giving in to their prejudices because of what the park means. This intersubjective group of people was influenced by what the park meant and signified for them, not only individually but also collectively. People saw and engaged with the walking routes, the trees carefully planted with space between them to enable those walking in the park to see each other, and the park benches at the end of the walking route so that people who encounter one another can sit and talk. This informed their behavior within the park, so that they were inclined toward openness, tolerance, and respect. The multistability of the park enabled people to create alternative stabilities for the park and turn it into a space that would take care of most of the basic needs of the protesters. Multiplicity and stability here play an essential role. Because of what the park allowed people to do, they were able to appropriate the park as a public space. Thus, the people who were in the park not only stopped its destruction but also created a community. The technology in the middle, in the move from the upper right curve to the lower left curve, enabled a human intentionality for *polis* making. However, as noted above, this intentionality would not be present without the first step of the technopolitical mediation, namely mediation of common sense.

The human, the technology, the world, and the intersubjective other all contribute to a new, temporary appearance of *praxis* in the world, which will later be stabilized within the common sense that people will share with their fellow humans, either in the form of cultural products or material objects. These cultural products and material objects will inform further appearances of *praxis* as well. What needs to be noted is that this two-step process of technopolitical mediation never comes to an end. It is always in motion, and what can be analyzed are the momentary, temporary mediations.

In the previous chapter, I argued that there was also a digital dimension that enabled people to participate in the Gezi Park protests. The role of social media platforms was unavoidable, in the sense that people were constantly informing one another from within and outside of the park. I also pointed out that the role of social media, as it was perceived by the respondents, tended to be conceptualized in terms of sharing information. Social media was especially influential in the first step of the technopolitical mediation, the mediation of common sense. In the section below, I delve into how social media mediated intersubjectivity among the participants of the protests. I do this with reference to the existing literature and the scripts of the interviews I conducted with the participants of the Gezi Park protests.

Technopolitical Mediation on Social Media

Now that I have outlined the two-step model of technopolitical mediation, I want to apply this model to understand the role of Facebook in the mediation of intersubjectivity. According to my respondents, the experience of political activity would be limited if it were only confined to digital spaces, like social media platforms or, specifically, Facebook. According to one of my respondents, political interactions are more effective when they are exercised in the form of face-to-face interaction:

> the valuable thing that we did in the park is that we were able to talk face-to-face. I mean, there is a difference between reading and seeing something and experiencing that thing, there is a difference between someone coming to you and saying pepper spray is very bad and you breathing that spray and breathing that for unfair reasons, there you see how bad it is; or someone telling you the righteousness of a movement with their tweets or telling you the same thing while looking into your eyes, because when you look into the eyes of people, you see how much they believe in the thing they are telling you, and if someone that is across from me is telling me this while believing in it, I have the urge to listen to them.

According to this respondent, an act is a political act only if it is exercised in public by people who have physical and bodily existence in the public space. It is not surprising that my respondent found face-to-face interaction more valuable than online interaction with regard to political experience. According to research conducted by Feezell et al., "in face-to-face interactions people are physically held accountable for their statements and conversation" (Feezell et al., 2009, p. 9), whereas on Facebook and other social media platforms, one can be anonymous and not held socially accountable for whatever content one shares. Gertz (2018) also observes that, as humans, we have a desire for "evading the burden of accountability" (Gertz, 2018, p. 9). Clicking on a link or sharing a piece of information allows us to feel justified; thus, platforms such as Facebook and Twitter contribute to this desire to avoid accountability (Gertz, 2018). Evgeny Morozov (2011) defines online political participation as "slacktivism," since he believes this form of participation gives one the feeling of being politically active through voicing opinions while not producing any political change or political impact. Morozov's analysis was shared by one of my respondents, when she defined her online political participation as "not playing by the book":

> What I do on Facebook, I'm giving an example, there was this Soma mine disaster, all of us changed our profile pictures to mines: that doesn't mean anything actually. Or, for example, we are making a show of mourning, like when Berkin Elvan died … who was a part of Gezi, we all made our profile pictures or we shared things about him, we followed news about him, but it doesn't exactly bring about the politicization of a person, it even makes you feel bad.

While some of my respondents, like the one quoted above, drew attention to physical or bodily existence and thus face-to-face interaction in the material space for politics to occur, another respondent perceived both digital media platforms and material spaces as equally valuable:

> imagine a gun … a gun divulges activity reciprocatively, it causes us to make different choices. In America, the gun lobby says humans kill not guns, let the use of guns be free. More left-wing people say no, because guns are deregulated and they are everywhere, that very easily changes people, thus guns also kill here. My position is closer to the second argument: I mean, having a tool does not bring about the use of that tool, only instrumentally; I believe that it has an impact, a transformative power on agency, I become a different person with my tool. Let's say you have a gun in your back pocket and it is empty, and I have a gun too but mine is loaded, now we are having a different discussion, do you understand what I mean? That's why Twitter is not a tool, Twitter is a realm … I mean the digital is not a tool for the original, it is just one of the manifestations of the original, I mean the realm on Twitter and the realm in the park are both real places that we actually exist, one is not the derivative or the tool of the other.

As the respondent states, it would be futile to disregard the influence of the two realms on one another in terms of political activity.

While the participants within the Gezi protests were also fragmented and polarized, Gezi allowed a space for interaction, inclusion, and debate for these people. However, there was another group of people in Turkey who were not at all present in the protests: AKP supporters. While the heterogeneous population within the park was able to find common ground and understand one another, Erdoğan supporters perceived the protesters as troublemakers, terrorists, and anarchists influenced by the harsh rhetoric of Erdoğan at the time (Traynor & Letsch, 2017). Unfortunately, while debate, discussion, openness, and understanding could overcome the polarization and fragmentation among the people within the park, the protesters failed to reach out to supporters of Erdoğan in terms of opening up spaces for discussion. What happened was rather as my respondent claimed:

> I believe in the power of social media, but it has its own limits, and we should act knowing these limits without living in the dream world … I agree that social media is a public realm but we should also see that it is a dangerous public realm because the polarization on the outside world can easily turn into, like, we play among one another … but if the logic of public space is being in interaction with other different perspectives who have different opinions than ours, especially in these times that Turkey is going through, that space gets tightened, but it is powerful in the manner of collecting people that are around you.

What this respondent seems to be referring to is the so-called filter bubble, a term coined by Eli Pariser (2011) to describe the situation whereby social media platforms hide from the user news, events, or opinion pieces that come from opposing sources, according to the pages, news, and events they follow. For Pariser, this leads to isolated people who only see or read information that like-minded people on their social media network see, and thus they live in a mode of reality where they do not see the other side of the coin (Pariser, 2011). It is true that this filter bubble is effective in preventing people from seeing what is posted and shared by those who are not like-minded. It is also true that when there is polarization and fragmentation in society, its reflection in social media platforms can result in even more vulgar confrontations. It would be foolish to disregard the influence of the digital realm on the physical realm and vice versa; however, it would also be wrong to treat social media platforms as a public space per se. I would like to highlight an important aspect of what my respondents said about Facebook. They focused on the physical or bodily experience of the political interaction as something that was an essential aspect of it, and they claimed that in the physical spaces, they were able to encounter people who were not like themselves. They could be exposed to others, whereas on social media, the situation is rather as my respondent claimed: "we play among one another." To more clearly illustrate Pariser's filter bubble, Zeynep Tufekci (2015) gives an example from her own experience as a Facebook user:

> In August 2014, I watched it all online as Facebook's news feed algorithm flooded my timeline with the Ice Bucket Challenge, a worthwhile charity drive for ALS research, while also algorithmically hiding updates from the growing Ferguson protests. Many of my friends were furiously posting about Ferguson on Facebook and on Twitter, but I only saw it on Twitter. Without Twitter to get around it, Facebook's news feed might have algorithmically buried the beginning of what has become a nationwide movement focusing on race, poverty and policing.
>
> (Tufekci, 2015)

Tufekci was able to navigate through other social media platforms to find news sources that informed her about a subject she would potentially be interested in. According to Gertz (2018), most of the time, we are not aware that the algorithms are working behind the scenes in our interactions with technologies. Referring to a recent study (Luckerson, 2015), Gertz argues that the researcher who tried to unravel people's online behavior was faced with the problem that 62 percent of people did not even know that their news feeds were being filtered (Gertz, 2018). Gertz, however, highlights a more ontological problem related to the news feed filters, which is our relationship with these algorithms: "What is at issue here is not that these algorithms claim to know us, but that we believe them" (Gertz, 2018, p. 96). Referring to Cass Sunstein (2007), Gertz argues that we have developed a belief that it is better for us to have tailored results when we are navigating the web. Gertz is critical of this "Daily Me," which is "a communications package that is personally designed, with each component fully chosen in advance" (Gertz, 2018, p. 98) to show us what to like and dislike. Yet he is more concerned about why we are drawn to the possibility of others delivering us what we might like or what people similar to us like. Gertz finds the answer to this question in conformism. As he argues,

> While Sunstein's worry is that such algorithmic curation is dangerous for the proper functioning of a democracy—as filtering leads to echo chambers, group

> polarization, extremism and "making America great again"—my interest here is rather in the question of what leads us to desire a "Daily Me" in the first place.
> (Gertz, 2018, p. 98)

One becomes comfortable among one's own crowd of people, without being challenged about identity or preferences, since there is an assumption that all the people in that crowd like the same things. As long as I do not see the other, as long as I am not confronted with an idea that I do not necessarily agree with, I have constant reassurance from my peers, my friends, and my news feed that they also see what I see. Does this Facebook page of mine, then, with my news feed, not seem to be closer to my private space, which I share with the company of my friends, than to a public space where I encounter others in the sense of encountering opposing, contradicting, and contesting views?

Assuming that one is technology-savvy like Tufekci (2015), who was able to inform herself about the Ferguson event, one can circumvent this filter bubble effect of Facebook. For example, as a Facebook user, I can follow Facebook groups that do not reflect my political ideas, worldview, or even everyday habits. In this way, I can inform myself more about what people think and share who are not necessarily my friends or within my close vicinity, insofar as they make that information publicly available. I can also circumvent this filter bubble or echo chamber (Flaxman et al., 2016) by increasing the number of pages and groups I join to obtain a truer picture of the world. Thus, it is possible to counter the claim that Facebook does not allow us to see opposing views by using the tools Facebook itself provides.

In the previous chapter, I showed that Facebook was influential in the protests by being a space for sharing information. Social media has been praised for being influential in generating access to various political and economic opportunities for everyone, facilitating collective action, and generating new forms of expression (Guess & Lyons, 2020). Facebook specifically has empowered many citizens in their demand for political change all over the world (Passini, 2012). For around five years after they were first launched, social media platforms were discussed in these celebratory terms. But in the second decade of the twentieth century, that began to change. Social media platforms like Facebook and Twitter started being discussed rather as hubs for "fake news and propaganda than as tools for empowerment and social change" (Guess & Lyons, 2020, p. 10). According to one of my respondents, the greatest difference between social media like Facebook and the park was

> Being able to touch people, I mean if the whole story is that now everybody has a story, that story starts with touching people, I couldn't do any of that on social media.

The respondent uses the expression "being able to touch people" metaphorically but also to highlight the value of physical nearness. This aspect of being bodily present in a space was conceptualized in relation to the Gezi Park protests by Zeynep Gambetti (2014) as well. According to Gambetti,

> Gezi was the struggle of disorderly bodies, of those who did not have any dispositive other than their bodies, against retaliatory machines. The words and signs were taken away from these bodies, but the bodies persisted in gathering in public space instead of dispersing or disappearing outright. The physical presence of the bodies performatively enacted the demands that were being silenced through material and immaterial means by the forces of the state. It was as if bodies were operating a kind of scission (or a decision) in a zone of indecision.
>
> (Gambetti, 2014, p. 96)

Gambetti also gives the example of the standing man, a performance artist who simply stood for eight hours on Taksim Square after the evacuation of the park, looking at the Atatürk Cultural Center directly behind Taksim Square (Gambetti, 2014). This act of presence as a form of resistance also inspired others who joined in the act of standing up, which "was enough to throw the state apparatus into disarray" (Gambetti, 2014, p. 97). Gambetti defines the protesters in the park as "living multiplicitly" (Gambetti, 2014, p. 98), highlighting the heterogeneity of their bodies in terms of identity, ideology, and religious belief. For Gambetti, the protesters' act of embodied resistance, their "thereness," defied problems that could arise from ideological differences (Gambetti, 2014, p. 99). Below, I illustrate how mediation of intersubjectivity was experienced by the participants of the Gezi Park protests.

Technopolitical Mediation on Facebook

The first step in the technopolitical mediation for the Facebook user is similar to that for the participants of the Gezi Park protests. A phenomenon or event has occurred in the world, which attracts their attention, and they want to understand the reason for the event or what the event they read or heard about was. Facebook is an important element in instigating the awakening of the user's attention. However, when the user whose attention has been awakened continues to have access to the world only through the social media platform, even the complementary aspect of this digital realm poses challenges for the existence of a plural public. When the Gezi protester moves from the digital realm and goes to the park itself, their access to the common sense is mediated by Gezi Park. However, if the user whose attention is awakened does not move beyond the digital realm—in our case, Facebook—Facebook sits in the way of access to common sense. As has already been seen, a public space—that is, a physical location—can be a reflection of the collective memory of the community, which in turn informs the understanding of the citizen. However, Facebook does not reflect the collective memory of the people as a society; it shows the user the information that the user is likely to want to see. In that sense, the process from preliminary understanding to explicit understanding does not refer to the common sense once the access to the world is mediated by Facebook; instead, it refers to what the user already believes in. If it is assumed that the user is technologically literate and is able to circumvent Facebook's algorithms, the access to the world may still be problematic, since the platform is a hub for disinformation. Being a space for possible diversion via disinformation and misinformation while people are trying to access their common sense, Facebook harbors the potential to manipulate the user. This move from the user to the world can be seen in the diagram of the lemniscate in Figure 3 (in the move from the upper left curve toward the lower right curve). This move for the Facebook user already produces a biased, misinformed, or disinformed citizen who is not challenged about their perspectives on the world but affirmed and confirmed in what they believe to be the truth. Either they are not exposed to other perspectives, or they have been disinformed about what the perspective of the other entails.

This brings me to the second step of technopolitical mediation, mediation of intersubjectivity. As previously mentioned, meaning making and the stabilization of meaning are collective activities. On Facebook, instead of being exposed to other people who are using the same platform differently than themself, the Facebook user is only exposed to their own community of like-minded people. Again, if it is assumed that the user is technologically literate and can access Facebook groups that are not like-minded, it is nevertheless the case that the first step of the technopolitical mediation has already been limited by the user's access to the world. Thus, the user is inclined to protect the group they belong to, rather than being open and understanding toward the other groups they encounter.

Like the technopolitical mediation in Gezi Park, multiplicity and stability play an essential role on Facebook as well. However, not only does the design of Facebook not allow multiple appropriations, but the user of Facebook also does not accept multiple appropriations. The technology in the middle, in the move from the upper right curve to the lower left curve in Figure 4, enables a herd mentality: a group in which everyone thinks the same, likes the same things, and expects one another to act the same. This not only limits the realm of plurality online, but it also limits the public realm offline.

By reflecting on the technopolitical mediation on Facebook, it can be concluded that whereas the technopolitical mediation in spaces with multistability fosters the manifestation of the authentic and free self, Facebook, since the design of the platform does not allow multiplicity of perspectives to be present, generates inauthentic selves that seek uniformity rather than multiplicity.

Conclusion

In this chapter, I have analyzed technological mediation of intersubjectivity. I explained what exactly intersubjectivity means according to Arendt's interpretation. Showing that Arendt was influenced by Heidegger's definition of *Dasein* while conceptualizing the political subject as authentic, I also showed that their philosophies differ in terms of how they relate the self to freedom and community. I showed that while Heidegger portrays togetherness as a threat to the manifestation of the authentic and free self, Arendt holds that community is a condition rather than a threat for a political subject. Based on this insight, I demonstrated that the manifestation of the authentic and free self is either encouraged or suppressed by the intersubjective relationships within the community in which one is engaged. Building on the distinctions established between Arendt and Heidegger's philosophies, I focused on the role of technology in these intersubjective relationships. I did this through an analysis of the existing literature and the interviews I conducted with participants of the Gezi Park protests in Istanbul. I illustrated that there is an intrinsic relationship between the authenticity or inauthenticity of the community and the material space that houses that community.

Following this analysis, I demonstrated that the technopolitical mediation of technologies is a two-stage process. I showed that Arendt's definition of public space (which requires plurality of perspectives and a stable material space) is inextricably linked to the concept of multistability: The more multistable a public space is, the more it invites several actors to visibly manifest their specific interpretations and to share a common space in their uniqueness and diversity. I showed that a community can be considered authentic when its members create new meaning by appropriating an artifact and allow different interpretations to coexist while denying the stabilized meaning of the artifact in question. In contrast, the inauthentic community would assume the stabilized meaning of an artifact while suppressing alternative interpretations and appropriations of the artifact in question. I showed that the organization of the demonstrators within the park during the protests was a specific interpretation of sharing a common space. This interpretation, however, was the process of a technopolitical mediation: The interaction between people and the material design of the park opened up new spaces for action in the world by encouraging the engagement of different subjectivities.

Finally, I provided an additional aspect of community that is missing in Kudina's (2019) hermeneutic lemniscate. I showed that the material space where people meet mediates both the way in which a community is established and the way in which one becomes a part of a community. Through this analysis, I showed that the manifestation of the authentic and free self that Arendt defines as the political subject does not only depend on cultural or institutional conditions but also on material conditions. The multistability of the park's design accumulates these stories and ultimately converts them into the common sense that engulfs the community and the space around it. Applying this insight to Facebook, I concluded that because the design of the platform does not allow for a variety of perspectives, it creates an inauthentic self that looks for uniformity rather than diversity.

Through technopolitical mediation of technologies, I showed that *poiesis*, by making the temporal appearance of *praxis* durable, allows *praxis* to be maintained and preserved within the world, thus contributing to our common sense. In addition to maintaining *praxis*, *poiesis*, by making the temporal appearance of *praxis* durable, instigates further appearances of *praxis* in the world. Gezi Park itself plays an active role in how it is going to be appropriated by *praxis* due to its design: Since the park is multistable by design, it allows itself to be appropriated in various ways. These appropriations, however, have to be visible to be maintained and transferred to further generations. Therefore, the appropriation by *praxis* is not an individual act but a collective one that enables the various members of a community simultaneously to participate in the appropriation of technology.

Notes

1 See Chapter 1, p. 47.

2 By claiming that there is no essence of a person, Arendt tries to strip action away from any causality. If it were accepted that there is an essence of a person, that would, to a certain extent, eliminate the influence and importance of life experiences that person shares with other people in the world. In addition, accepting that there is a human essence would make human action a form of behavior that can be foreseen, monitored, and even controlled. Thus, by defining human action as *praxis*, Arendt denies that there is a human essence (Villa, 1995).

3 See pp. 59–61.

4 *Dasein*, first and foremost, is being toward death. As can be seen in the paragraphs above, it is also within *Dasein* to forget that it is toward death. Only through anxiety—the existential *Angst*—does *Dasein* wake up from its everyday tranquility. There is a vast amount of literature that covers the Kierkegaardian roots of Heidegger's existentialism in his analysis of *Dasein* (Buben, 2013; Dreyfus & Rubin, 1987; Hall, 1984; Yagi, 2009). For the purposes of this book, it is fruitful to explain that Arendt is opposed to Heidegger's characterization of *Dasein*'s focus on death. Whereas Heidegger posits *Dasein* as being forced to confront the lack of meaning when facing death, therefore starting the process of authentic disclosing of the world to generate meaning and itself, Arendt positions *natality* as the initial moment of the process of authentic disclosing. Influenced by St. Augustine, Arendt focuses on the meaning-creating capacity of the newborn or new life. That is why and how Arendt is able to position "action" or "praxis" as being like a second birth. In that sense, Arendt is able to dismiss the ontological focus of *Dasein* on itself and its own death and move toward the others (Villa, 1995, pp. 132–5).

5 See Chapter 2, pp. 92–8.

6 Platform of Gezi Park (my translation).

7 The respondent uses the term "politics" here in the sense of being discriminatory toward others in relation to one's own political ideologies. This use of the term became clear from the follow-up questioning.

8 At the time of the interview, the third bridge on the Bosphorus, connecting the Asian and European parts of Istanbul, was under construction.

9 Yavuz Sultan Selim was the Ottoman Sultan who ruled the empire between 1512 and 1520, well known by the Alevite community because of his massacres of Alevites to punish their cooperation with the Persian Safavids (Üşenmez & Duman, 2015). In the follow-up questioning, the respondent expressed his dissent as to the choice of the name for the bridge, since to him, naming a bridge after a sultan who massacred Alevites shows sectarianism, favoring one sect of the religious population over the other.

10 See Chapter 3, pp. 119–22.

Conclusion

In this concluding chapter, I would like to first return to the main research question of this book and reflect on the extent to which it has been answered. Second, I return to the canonical case study that inspired this book and which I have criticized in the introduction: the case of the Moses Bridges as analyzed by Langdon Winner (Winner, 1980). In the third section, I sketch directions for further research—for instance, how the findings of this book can be expanded to apply to research into social media or artificial intelligence (AI).

In this book, I asked a question: *How do technologies inform the way in which political communities are formed by mediating our political interpretations and interactions?*

During the quest to answer this question, I have shown that the politics of technology could be better understood when we move beyond a simplistic view that focuses solely on power relations (as seen in Winner's and Feenberg's approaches). I showed that this traditional approach is not sufficient since it is based only on one aspect of power, which is *power over*, and consequently reduces politics to hierarchical control. I have shown that the various concepts of power, whether "power over" or "power to," are fundamentally relational, requiring interaction with others to exist and be understood, thus emphasizing that power's essence lies in human relationships that shape shared reality (p. 15). This allowed me to position interaction rather than power over as foundational to politics. So I have shown that to understand the political implications of technologies, we need to move beyond an analysis of how technologies exert power over people and analyze how technology facilitates interaction and shapes the formation of political communities.

For that purpose, integrating Arendt's political theory, which focuses on the interaction between free individuals in a public space (the polis), with technological mediation theory, I have shown that the political impact of technologies is best understood in unpacking (a) how they shape common interpretative frameworks (technological mediation of common sense) and (b) how they structure intersubjective dynamics (technological mediation of intersubjectivity). The technopolitical mediation developed in this book acknowledges and reinstates the complex and relational aspect of politics. Technopolitical mediation acknowledges a dynamic coproduction of shared meaning and interaction between humans, technologies, and material worlds. Thus, technopolitical mediation shows us that technologies condition, frame, and organize politics.

Revisiting the Moses Bridges with Technopolitical Mediation

As I mentioned in the introduction, Langdon Winner, in his influential article "Do Artifacts Have Politics?" (Winner, 1980) gives the example of the low-hanging bridges on Long Island, New York, designed by Robert Moses. Because they were so low, the bridges did not allow public transportation vehicles, such as buses (which are around 9 feet tall), to pass underneath them. According to Winner, Moses inscribed his ideas about social class and racial prejudice in the artifacts he designed (Winner, 1980). On the basis of this—and other—examples, Winner claims that technologies can be inherently political and can embody ideologies and "specific forms of power and authority" (Winner, 1980, p. 121). Moses' bridges are "political," according to Winner (1980), because they carry out the racially biased ideology of their designer.

This famous example helps to illustrate what I have been arguing: that the political dimension of technological artifacts is primarily analyzed in terms of power. Power is also essential in Winner's definition of politics as "arrangements of power and authority in human associations as well as the activities that take place within those arrangements" (Winner, 1980, p. 123). Power, in the way in which Winner (1980) characterizes it, can be interpreted as *power over*. As I have shown, *power over* always implies and presupposes another source of power, and this means that power is always relational. In that sense, power can only be conceived of within a relationship, since one always needs another to articulate or to understand what power is. Thus, interaction is already inscribed in the notion of power. What kind of an analysis of the impact of technology on politics, then, would result from moving beyond *power over* as the basis of politics and rereading this example with this newly expanded conception of the politics of technology in mind?

As I have argued throughout this book, to grasp the political significance of technology, it is necessary to go through a two-step analysis of the technopolitical mediation: first, the mediation of common sense, and second, the mediation of intersubjectivity. If the framework of technopolitical mediation is applied to the example of the Moses Bridges and their impact on society, what kind of new insights can this yield, parallel to Winner's (1980) analysis?

Applying the technopolitical mediation to Winner's (1980) example of Moses' bridges, in the move from the upper left curve to the lower right, the first step of the technopolitical mediation can be observed. In this first step, it can be seen that the bridges amplify the segregation of the people. Segregation has a long history in the United States,[1] and thus it has a place in the common sense of the people of New York as well. In that sense, Moses' bridges, by actively preventing access only to people relying on public transport (coincidentally people of moderate income or with a particular skin color), reminded the people of New York of what they already knew because of their cultural memory. While some authors (Joerges, 1999; Woolgar & Cooper, 1999) have argued that people did not necessarily have to use the highway to reach the beaches on Long Island, that does not change what this technology signifies. Even though it was possible to reach the beaches of Long Island by other means, the fact that these bridges actively stopped a certain group of people from reaching a public space reminded the people of New York of segregation and of what it means to be deprived of freedom. Bridges appeal to part of the cultural memory of New York, and they are a material form of this cultural memory of segregation. The bridges segregated society and had the power to answer the question of who was allowed to access the beaches of Long Island.

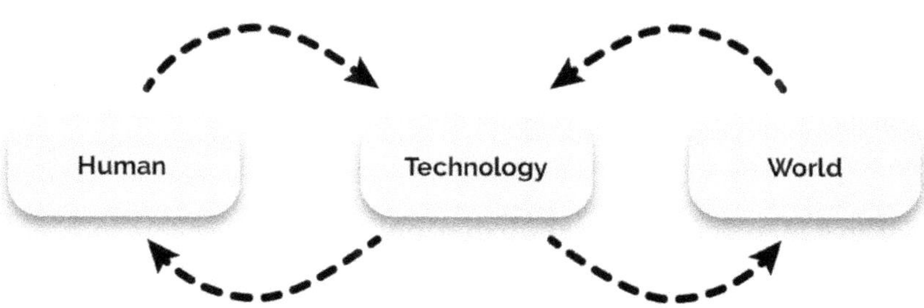

Figure 2 The Hermeneutic Lemniscate.

This leads to the second step of the technopolitical mediation, the mediation of intersubjectivity. In the move from the world, through technology, to humans, it can be seen that the bridges not only prevented buses from reaching the beach but also revealed a divide within the people of New York. The bridges indirectly continued to instigate discussions about this divide. By bringing segregation to the fore, Moses' bridges framed people who take the bus as people without enough wealth, or with a particular ethnic background, who should be excluded from specific parts of public space—and in doing so, the bridges made this framing available for political interaction. The intersubjective relations between humans had already been conditioned by this technology in framing one group as the powerful and another group as the powerless. The bridge provided a narrative—a context, and content for the interaction between these two groups. As such, the respective sources of power that the richer and poorer citizens of New York had were not equal: One was above the other. In this political engagement, there was limited room for plurality of perspectives or even for negotiating positions among different perspectives. Thus, the intersubjective relationships between humans already had a certain conditioning and framing surrounding them.

As can be seen, the bridges organized political interactions between people in a *material* way, which in turn influences the world-disclosing capacity of technologies that relate human beings to each other in *interpretive* ways. In this example, then, technology not only exerts *power over* by preventing one specific group in society from reaching a public space, but it also influences the interpretative framework of people in the ways in which it sheds light on the common cultural history of the country in which it is embedded. In that way, it conditions the discussions among subjects by framing them as "the powerful who have access to Jones Beach" and "the powerless who need to use public transport."

The technopolitical mediation approach is not only a sophisticated intellectual accomplishment but it can also be a conceptual framework that can be applied to different digital platforms that shape digital society and inform scholars about the desirability of the emerging technopolitical mediations.

The Future of Technopolitical Mediation

In this book, I have extensively built upon the interdependent relationship between Arendtian distinctions of *poiesis* (world building) and *praxis* (world making). I have shown that the end products of *poiesis*, whether cultural or material, stabilize the appearances of *praxis*. In doing so, *poiesis* also inspires further appearances of *praxis*. Products of *poiesis* are an embodiment of the meaning that the world has been given at their time of creation. However, to remain meaningful for people, these products constantly need to be appropriated and perceived as meaningful by *praxis*. Thus, the stabilization performed by *poiesis* is dependent on *praxis*'s appropriation of it.

Earlier, I gave the example of a sculptor as a spectator of the Hungarian Revolution of 1956. This person, I indicated, wants to turn the temporal appearance of the revolutionaries into something stable, and so he creates a sculpture. After its creation, what this artwork *means* is based on its appropriation through *praxis*, since even though *homo faber* as a spectator was the initial creator of the sculpture as a testimony to the appearance of *praxis*, the meaning that is going to be ascribed to that sculpture cannot be fixed by *homo faber*. Now that the sculpture is out in the world, it is an appearance that will be perceived by the acting agent, and its meaning will depend on the hermeneutic circle that will form once the agent engages with the sculpture. In other words, it is the interpretative act, as well as the act of creation, that prospectively creates durability. However, I have not elaborated on the fact that *poiesis*, and the context in which *poiesis* is going to stabilize *praxis*'s appearance, is also dependent on the medium that *poiesis* utilizes for this stabilization. The way in which this sculpture stabilizes the appearance of *praxis* is, to an extent, dependent on the materials that were utilized for its creation: The wood or the marble, the chisels—all are influential in the final shape of the sculpture. This, in turn, influences the way in which *praxis* further appropriates *poiesis*. A statue would not be appropriated in the same way as a painting, a story, or a public park. In that sense, reinterpreting *poiesis* depends on other media that enable retrieving its meaning and appropriating it in a particular way. For instance, technopolitical mediation on social media platforms such as Reddit and 4chan differs from what I have presented in this book. During the Gezi Park protests, the material space and technological arrangement of the park fostered political engagement and allowed for the manifestation of an authentically free self while generating an authentic political community. However, if we consider the role of social media platforms, such as Twitter (now X), Reddit, and 4chan, in the creation and evolution of the Pizzagate conspiracy theory (Wendling, 2016), we cannot discuss the generation of an authentic political community. In our current platform-dominated digital environment, the infrastructure is specifically curated to generate clusters of people who share the same beliefs without being exposed to alternative perspectives. The spread of Pizzagate indicated how platforms, including 4chan and Reddit, were essential for mediating the intersubjective relationships between individuals, enabling them to shape and reinforce each other's beliefs. Through repeated, coordinated performances, memes, hashtag campaigns, and videos, people circulated, validated, and reaffirmed each other's beliefs. This online intersubjectivity resulted in a new worldview, generating a common distrust in which the boundaries of truth and falsity were continuously blurred. Thus, online-generated conspiracy theories such as Pizzagate can be analyzed with technopolitical mediation as interpretative artifacts that mediate political identities and generate clusters of solidarity among fragmented publics.

Examples of online intersubjectivity, however, do not necessarily produce misinformation or conspiracy theories. Locally bound, technologically driven issues can also be spread worldwide, creating a common understanding inclusive of differing perspectives and mutual trust. For example, the locally bound case of women exposing sexual harassment in the workplace in the United States with the hashtag *#MeToo* found a receptive audience in France, Spain, Italy, the UK, and Sweden, leading to legislative changes in some cases (Carbonaro, 2022). Technopolitical mediation in this case can help analyze how algorithmic filtering mediates collective interpretations of events such as *#MeToo*.

Moreover, our platform-driven digital environment is becoming increasingly AI-driven. As I write this, the world is talking about new large language models (LLMs), DeepSeek V3 and DeepSeek R1, owned by a Chinese company and released January 20, 2025. Like any other LLM, they are machine learning models trained with immense data to develop the capacity to understand and generate natural language and a wide range of content (IBM, 2023). DeepSeek V3 and DeepSeek R1 are AI-based chatbots, similar to ChatGPT from OpenAI and Gemini from Google. What was surprising about these chatbots was that they had math, coding, and reasoning abilities comparable to those of the subscription version of OpenAI's chatbot o1 but were produced at 90–95 percent less cost. In addition, users do not need to pay to access them (Booth, 2025). While some researchers perceive the political impact of such AI chatbots as the spread of misinformation (Chen et al., 2021; Weidinger et al., 2021) or political deskilling by hindering the capacity of individuals for critical engagement and collective action (Kudina & de Boer, 2024), technopolitical mediation can help extend these analyses by exploring how chatbots actively shape people's interpretative frameworks in the first place. Technopolitical mediation can help analyze how

LLMs homogenize narratives by generating "neutral" summaries that obscure marginalized perspectives while restructuring shared understandings of political issues. In summary, further research on the political impact of technologies should focus on digital public spaces since these are the new media through which collective meanings and political interpretations are stabilized.

The approach that is developed in this book enables an expansion and a deeper understanding of the political impact of technologies. Moving beyond the discussion of power, the technopolitical mediation approach shows how technologies condition, frame, instigate, and organize politics. Artifacts have politics not only because they embody power over one group of people but because they relate human beings to each other in specific ways. They do this by conditioning public spaces for political interaction and by revealing the world in specific ways as the basis for this political interaction.

Note

1 Segregation has been an important historical aspect of American social and political life. It is not within the scope of this book to discuss in detail the impact of segregation on the American population. For more information, see Massey & Denton (1993).

References

Acar, Y. G., & Uluğ, Ö. M. (2014). *Bir olmadan biz olmak: Farkli Gruplardan aktivistlerin Gözüyle Gezi direnisi.* Ankara, Turkey: Dipnot Yayinlari.

Ahmad, F. (1993). *The making of modern Turkey.* London, England: Routledge.

Akpınar, I., & Şoher, Ş. (2013, October). *«GEZİ»den sonra Kültür Vadisine bakmak: Yeni Yaşam ve Tasarım Rehberi mümkün mü.* GEZİ DEN SONRA TAKSİM konferansı, Istanbul, Turkey. https://www.researchgate.net/publication/276269227_GEZIden_sonra_Kultur_Vadisine_bakmak_Yeni_Yasam_ve_Tasarim_Rehberi_mumkun_mu

Arda, B. (2015). Apolitical is political: An ethnographic study on the public sphere in the Gezi uprising in Turkey. *Interface, 7*(1), 9–18. http://www.interfacejournal.net/wordpress/wpcontent/uploads/2015/06/Issue-7-1-Arda.pdf

Arendt, H. (1958). Totalitarian imperialism: Reflections on the Hungarian revolution. *The Journal of Politics, 20*(1), 5–43. https://doi.org/10.2307/2127387

Arendt, H. (1959). Reflections on Little Rock. *Dissent, 6*(1), 45–56.

Arendt, H. (1960a). Freedom and politics: A lecture. *Chicago Review, 14*(1), 28–46. https://www.jstor.org/stable/i25733547

Arendt, H. (1960b). Society and culture. *Daedalus, 89*(2), 278–87. http://www.jstor.org/stable/20026571

Arendt, H. (1970). *On violence.* Boston, MA: Houghton Mifflin Harcourt.

Arendt, H. (1977a). *The life of the mind.* Vol. *1*: *Thinking.* New York, NY: Harcourt Brace Jovanovich (LM).

Arendt, H. (1981a). *The life of the mind* (Vols. *1–2*). New York, NY: Harcourt Brace Jovanovich.

Arendt, H. (1981b). *Vita activa oder vom tätigen Leben.* München, Germany: Piper.

Arendt, H. (1982). *Lectures on Kant's political philosophy.* Chicago, IL: University of Chicago Press.

Arendt, H. (1990). Philosophy and politics. *Social Research, 57*(1), 73–103. https://www.jstor.org/stable/40970579

Arendt, H. (1998). *The human condition* (2nd ed.). Chicago, IL: University of Chicago Press (Original work published 1958).

Arendt, H. (2003). Personal responsibility under dictatorship. In J. Kohn (Ed.), *Responsibility and judgment* (pp. 17–49). New York, NY: Schocken Books.

Arendt, H. (2004). *The origins of totalitarianism.* New York, NY: Harcourt Brace Jovanovich.

Arendt, H. (2006a). *Between past and future: Eight exercises in political thought.* New York, NY: Penguin Books.

Arendt, H. (2006b). *On revolution*. New York, NY: Penguin Books.

Arendt, H., & Kohn, J. (Eds.) (2005a). *Essays in understanding, 1930–1954: Formation, exile, and totalitarianism*. New York, NY: Schocken Books.

Arendt, H., & Kohn, J. (Eds.) (2005b). *The promise of politics*. New York, NY: Schocken Books.

Aydin, C. (2009). The struggle between ideals: Nietzsche, Schmitt and Lefort on the politics of the future. In H. Siemens & V. Roodt (Eds.), *Nietzsche, power and politics* (pp. 801–16). Berlin, Germany: De Gruyter.

Baban, E., & Güzel, E. (2015). Digital activism and social movements: How the world perceives the Gezi movement. *Amity Journal of Media & Communications Studies, 5*(1), 16–26.

Balaban, O. (1990). Praxis and poesis in Aristotle's practical philosophy. *The Journal of Value Inquiry, 24*(3), 185–98. https://doi.org/10.1007/BF00149432

Baugh, B. (1989). Heidegger on Befindlichkeit. *Journal of the British Society for Phenomenology, 20*(2), 124–35.

Baykan, A., & Hatuka, T. (2010). Politics and culture in the making of public space: Taksim Square, 1 May 1977, Istanbul. *Planning Perspectives, 25*(1), 49–68.

Bénatouïl, T., & Bonazzi, M. (2012). Theoria and bios theoretikos from the Presocratics to the end of antiquity: An overview. In T. Bénatouïl & M. Bonazzi (Eds.), *Theoria, praxis, and the contemplative life after Plato and Aristotle* (pp. 1–14). Leiden, Netherlands: Brill.

Berkowitz, R. (2014). Did Eichmann think? *The Good Society, 23*(2), 193–205. https://doi.org/10.5325/goodsociety.23.2.0193

Bernstein, R. J. (1986). Rethinking the social and the political. *Graduate Faculty Philosophy Journal, 11*(1), 111–30.

Bilal, B., & Nahrwold, J. J. (2013). Urban utopia: A feminist perspective on the Gezi resistance. *Open Citizenship, 4*(2), 88–93.

Booth, C. (2025, January 27). *DeepSeek is breaking the internet*. Forbes. https://www.forbes.com/sites/callumbooth/2025/01/27/deepseek-is-breaking-the-internet/

Borren, M. (2010). *Amor mundi: Hannah Arendt's political phenomenology of world*. Amsterdam, The Netherlands: F & N Eigen Beheer.

Borren, M. (2013). A sense of the world: Hannah Arendt's hermeneutic phenomenology of common sense. *International Journal of Philosophical Studies, 21*(2), 225–55.

Braun, M. (1994). *Hannah Arendts transzendentaler Tätigkeitsbegriff: Systematische Rekonstruktion ihrer politischen Philosophie im Blick auf Jaspers und Heidegger*. Bern, Switzerland: Peter Lang.

Buben, A. (2013). Heidegger's reception of Kierkegaard: The existential philosophy of death. *British Journal for the History of Philosophy, 21*(5), 967–88. https://doi.org/10.1080/09608788.2013.825576

Canovan, M. (1983). Arendt, Rousseau, and human plurality in politics. *The Journal of Politics, 45*(2), 286–302.

Canovan, M. (1994). *Hannah Arendt: A reinterpretation of her political thought.* Cambridge, England: Cambridge University Press.

Carbonaro, G. (2022, October 13). *#MeToo 5 years on: What has changed in Europe since the start of the movement?* Euronews. https://www.euronews.com/culture/2022/10/08/metoo-5-years-on-what-has-changed-in-europe-since-the-start-of-the-movement

Carr, W. (2006). Philosophy, methodology and action research. *Journal of Philosophy of Education*, *40*(4), 421–35. https://doi.org/10.1111/j.1467-9752.2006.00517.x

Cats-Baril, W. L., & Jelassi, T. (1994). The French videotex system minitel: A successful implementation of a national information technology infrastructure. *MiS Quarterly*, *18*(1), 1–20.

Cavarero, A. (2002). Politicizing theory. *Political Theory*, *30*(4), 506–32.

Chen, M., Tworek, J., Jun, H., Yuan, Q., Pinto, H. P. D. O., Kaplan, J., Edwards, H., Burda, Y., Joseph, N., Brockman, G., Ray, A., Puri, R., Krueger, G., Petrov, M., Khlaaf, H., Sastry, G., Mishkin, P., Chan, B., Gray, S., & Zaremba, W. (2021). *Evaluating large language models trained on code.* arXiv. https://doi.org/10.48550/arXiv.2107.03374

Christman, J. (2020). Autonomy in moral and political philosophy. In Edward N. Zalta (Ed.), *The Stanford encyclopedia of philosophy* (Fall 2020 Edition). Stanford, CA: Metaphysics Research Lab, Stanford University.

Chrona, S., & Bee, C. (2017). Right to public space and right to democracy: The role of social media in Gezi Park. *Research and Policy on Turkey*, *2*(1), 49–61. https://doi.org/10.1080/23760818.2016.1272267

Cohen, A. J. (1994). Reviewed work: Critical theory of technology by Andrew Feenberg. *Theory and Society*, *23*(4), 575–82. http://www.jstor.org/stable/657892

Cohen, Z., Lyngaas, S., & Murray, S. (2024, October 31). *Election officials are outmatched by Elon Musk's misinformation machine | CNN politics.* CNN. https://edition.cnn.com/2024/10/31/politics/election-officials-outmatched-elon-musk-misinformation/index.html

Dahl, R. A. (1957). The concept of power. *Behavioral Science*, *2*(3), 201–15.

Demirhan, K. (2014). Social media effects on the Gezi Park movement in Turkey: Politics under hashtags. In B. Patrut & M. Patrut (Eds.), *Social media in politics* (pp. 281–314). Cham, Switzerland: Springer.

Disch, L. J. (1993). More truth than fact: Storytelling as critical understanding in the writings of Hannah Arendt. *Political Theory*, *21*(4), 665–94. https://doi.org/10.1177/0090591793021004006

Doppelt, G. (2001). What sort of ethics does technology require? *The Journal of Ethics*, *5*(2), 155–75. http://www.jstor.org/stable/25115687

Dossa, S. (1989). *The public realm and the public self: The political theory of Hannah Arendt.* Waterloo, Canada: Wilfrid Laurier University Press.

Dreyfus, H. L., & Rubin, J. (1987). You can't get something for nothing: Kierkegaard and Heidegger on how not to overcome nihilism. *Inquiry*, *30*(1–2), 33–75. https://doi.org/10.1080/00201748708602110

Eifert, S. (2024, July 25). *Photo of Kamala Harris embracing Jeffrey Epstein on beach is fake*. Snopes. https://www.snopes.com/fact-check/kamala-harris-jeffrey-epstein-beach-photo/

Eken, B. (2014). The politics of the Gezi Park resistance: Against memory and identity. *South Atlantic Quarterly*, *113*(2), 427–36. https://doi.org/10.1215/00382876-2644212

Elpidorou, A., & Freeman, L. (2015). Affectivity in Heidegger I: Moods and emotions in being and time. *Philosophy Compass*, *10*(10), 661–71. https://doi.org/10.1111/phc3.12236

Ete, H. (2013). The political reverberations of the Gezi protests. *Insight Turkey*, *15*(3), 15–25.

Feenberg, A. (1992). Subversive rationalization: Technology, power, and democracy. *Inquiry*, *35*(3–4), 301–22. https://doi.org/10.1080/00201749208602296

Feenberg, A. (1995). *Alternative modernity: The technical turn in philosophy and social theory*. Berkeley, CA: University of California Press.

Feenberg, A. (1999). *Questioning technology*. Oxfordshire, England: Routledge.

Feenberg, A. (2008). *Encountering technology* [speech]. Retrieved from http://www.sfu.ca/~andrewf/encountering.pdf

Feenberg, A. (2009). Peter-Paul Verbeek: Review of what things do. *Human Studies*, *32*(2), 225–8. https://doi.org/10.1007/S10746-009-9115-3

Feenberg, A. (2010). *Between reason and experience: Essays in technology and modernity*. Cambridge, MA: MIT Press.

Feenberg, A. (2013). What I said and what I should have said: On critical theory of technology. *Techne: Research in Philosophy and Technology*, *17*(1), 163–78. https://doi.org/10.5840/techne20131719

Feezell, J. T., Conroy, M., & Guerrero, M. (2009). Facebook is … fostering political engagement: A study of online social networking groups and offline participation. Presented at Annu. Meet. Am. Polit. Sci. Assoc., September 3–6, Toronto. http://papers.ssrn.com/sol3/papers.cfm?abstract_id=1451456&rec=1&srcabs=1331029

Flaxman, S., Goel, S., & Rao, J. M. (2016). Filter bubbles, echo chambers, and online news consumption. *Public Opinion Quarterly*, *80*(1), 298–320. https://doi.org/10.1093/POQ/NFW006

Freedom House. (2014). *Freedom of the press*. New York, NY: Freedom House.

Gambetti, Z. (2014). Occupy Gezi as politics of the body. In U. Özkırımlı (Ed.), *The making of a protest movement in Turkey #occupygezi* (pp. 89–102). New York, NY: Springer.

Gertz, N. (2018). *Nihilism and technology*. Lanham, MD: Rowman & Littlefield International.

Gilbert, D. (2024, October 21). *Russian propaganda unit appears to be behind spread of false Tim Walz sexual abuse claims.* WIRED. https://www.wired.com/story/russian-propaganda-unit-storm-1516-false-tim-walz-sexual-abuse-claims/

Goehler, G. (2000). Constitution and use of power. In H. Goverde, P. G. Cerny, M. Haugaard, & H. H. Lentner (Eds.), *Power in contemporary politics* (pp. 40–68). Thousand Oaks, CA: Sage.

Göhler, G. (2009). Power to and power over. In S. R. Clegg & M. Haugaard (Eds.), *The Sage handbook of power* (pp. 27–37). Thousand Oaks, CA: Sage.

Göle, N. (2013a). The Gezi occupation: For a democracy of public spaces. Retrieved from https://www.opendemocracy.net/en/gezi-occupation-for-democracy-of-public-spaces/

Göle, N. (2013b). Gezi-anatomy of a public square movement. *Insight Turkey*, *15*(3), 7–14.

Göle, N. (2013c). Yer Sofrası ve Sınır İhlalleri. Retrieved from https://t24.com.tr/yazarlar/nilufer-gole/yer-sofrasi-ve-sinir-ihlalleri,7056

Guess, A. M., & Lyons, B. A. (2020). Misinformation, disinformation, and online propaganda. In N. Persily & J. A. Tucker (Eds.), *Social media and democracy: The state of the field, prospects for reform* (pp. 10–33). Cambridge, United Kingdom; New York, NY: Cambridge University Press.

Gül, M. (2009). *Emergence of modern Istanbul: Transformation and modernisation of a city* (Vol. *83*). London, England: IB Tauris.

Gül, M., Dee, J., & Nur Cünük, C. (2014). Istanbul's Taksim Square and Gezi Park: The place of protest and the ideology of place. *Journal of Architecture and Urbanism*, *38*(1), 63–72. https://doi.org/10.3846/20297955.2014.902185

Gürcan, E., & Peker, E. (2015). *Challenging neoliberalism at Turkey's Gezi Park: From private discontent to collective class action*. New York, NY: Springer.

Haciyakupoglu, G., & Zhang, W. (2015). Social media and trust during the Gezi protests in Turkey. *Journal of Computer-Mediated Communication*, *20*(4), 450–66. https://doi.org/10.1111/jcc4.12121

Hall, H. (1984). Love and death: Kierkegaard and Heidegger on authentic and inauthentic human existence. *Inquiry*, *27*(1–4), 179–97. https://doi.org/10.1080/00201748408602017

Hansen, P. B. (1993). *Hannah Arendt: Politics, history and citizenship*. Hoboken, NJ: Wiley.

Harman, G. (2014). *Bruno Latour: Reassembling the political*. London, England: Pluto Press.

Haugaard, M. (2012). Rethinking the four dimensions of power: Domination and empowerment. *Journal of Political Power*, *5*(1), 33–54. https://doi.org/10.1080/2158379X.2012.660810

Heidegger, M. (1954). *The question concerning technology*. Retrieved from http://www2.hawaii.edu/~freeman/courses/phil394/The%20Question%20Concerning%20Technology.pdf

Heidegger, M. (1997). *Kant and the problem of metaphysics*. Bloomington, IN: Indiana University Press.

Heidegger, M., Macquarrie, J., & Robinson, E. (1962). *Being and time*. Oxford, England: Blackwell.

Hill, M. A. (Ed.) (1979). *Hannah Arendt: The recovery of the public world*. New York, NY: St Martin's Press.

Honig, B. (1993). *Political theory and the displacement of politics*. Ithaca, NY: Cornell University Press.

Hudnall, H. (2024, October 3). *Image of Donald Trump wading through flood water is AI-generated | Fact check*. https://eu.usatoday.com/.https://eu.usatoday.com/story/news/factcheck/2024/10/03/ai-image-trump-hurricane-helene-fact-check/75483588007/

IBM. (2023, November 2). *What are large language models (LLMs)?* https://www.ibm.com/think/topics/large-language-models

Ihde, D. (1990). *Technology and the lifeworld: From garden to earth*. Bloomington, IN: Indiana University Press.

Ihde, D. (1993). *Postphenomenology: Essays in the postmodern context*. Evanston, IL: Northwestern University Press.

Ihde, D. (1998). *Expanding hermeneutics: Visualism in science*. Evanston, IL: Northwestern University Press.

Ihde, D. (1999). Technology and prognostic predicaments. *AI & Society*, *13*(1), 44–51. https://doi.org/10.1007/BF01205256

Ihde, D. (2009). *Postphenomenology and technoscience: The Peking University lectures*. Albany, NY: State University of New York Press.

İnsel, A. (2013). Haysiyet ayaklanması. *Revolt of Dignity, Radikal*, *4*(1), 19.

Inwood, M. (1999). *A Heidegger dictionary*. Hoboken, NJ: Wiley.

Jandrić, P. (2019). Review of Eduardo Beira and Andrew Feenberg (Eds.) (2018). Technology, modernity and democracy: Essays by Andrew Feenberg. *Postdigital Science and Education*, *1*(2), 551–7.

Jaspers, K. (2010). *Man in the modern age*. London: Routledge.

Joerges, B. (1999). Do politics have artefacts? *Social Studies of Science*, *29*(3), 411–31.

Kaplan, D. M. (2009). What things still don't do. *Human Studies*, *32*(2), 229–40. https://doi.org/10.1007/S10746-009-9116-2

Karakayalí, S., & Yaka, Ö. (2014). The spirit of Gezi: The recomposition of political subjectivities in Turkey. *New Formations*, *83*(83), 117–38. https://doi.org/10.3898/NEWf.83.07.2014

Kateb, G. (2000). Political action: Its nature and advantages. In D. R. Villa (Ed.), *The Cambridge companion to Hannah Arendt* (pp. 130–48). Cambridge, England: Cambridge University Press.

Kaya, A. (2015). Islamisation of Turkey under the AKP rule: Empowering family, faith and charity. *South European Society and Politics*, *20*(1), 47–69. https://doi.org/10.1080/13608746.2014.979031

Kelz, R. (2016). *The non-sovereign self, responsibility, and otherness: Hannah Arendt, Judith Butler, and Stanley Cavell on moral philosophy and political agency*. New York, NY: Springer.

Keyder, Ç. (2014). Yeni Orta Sınıf. Retrieved on October 13, 2015 from https://acikradyo.com.tr/arsiv-icerigi/yeni-orta-sinifa-dikkat

Knauer, J. T. (1980). Motive and goal in Hannah Arendt's concept of political action. *American Political Science Review*, *74*(3), 721–33.

Konda Araştırma ve Danışmanlık. Gezi Raporu. Toplumun "Gezi Parkı Olayları" Algısı Gezi Parkındakiler Kimlerdi? 2014. http://www.konda.com.tr/tr/raporlar/KONDA_GeziRaporu2014.pdf

Kudina, O. (2019). *The technological mediation of morality: Value dynamism, and the complex interaction between ethics and technology* [Dissertation]. Retrieved from https://ris.utwente.nl/ws/portalfiles/portal/112667674/Olya_Kudina_2019_The_Technological_Mediation_of_Morality_PhD_Thesis.pdf

Kudina, O., & de Boer, B. (2024). Large language models, politics, and the functionalization of language. *AI and Ethics*, *5*(3), 1–13. https://doi.org/10.1007/s43681-024-00564-w

Kuymulu, M. B. (2013). Reclaiming the right to the city: Reflections on the urban uprisings in Turkey. *City*, *17*(3), 274–78. https://doi.org/10.1080/13604813.2013.815450

Latour, B. (1987). *Science in action: How to follow scientists and engineers through society*. Cambridge, MA: Harvard University Press.

Latour, B. (1994). Pragmatogonies: A mythical account of how humans and nonhumans swap properties. *American Behavioral Scientist*, *37*(6), 791–808.

Latour, B. (2005). From realpolitik to dingpolitik. In B. Latour & P. Weibel (Eds.), *Making things public: Atmospheres of democracy* (pp. 14–41). Cambridge, MA: MIT Press.

Latour, B. (2007). Turning around politics: A note on Gerard de Vries' paper. *Social Studies of Science*, *37*(5), 811–20.

Leonard, S. T. (2019). Introduction: A genealogy of the politicized intellectual. In D. Reid, S. Leonard, & L. Fink (Eds.), *Intellectuals and public life* (pp. 1–29). Ithaca, NY: Cornell University Press.

Letsch, C. (2013, June 3). *Social media and opposition to blame for protests, says Turkish PM*. The Guardian (Istanbul). Available at: http://www.theguardian.com/world/2013/jun/02/turkish-protesterscontrol-istanbul-square (Accessed: January 10, 2018).

Lewis, B. (1968). *The emergence of modern Turkey*. Oxford: Oxford University Press.

Lindqvist, E., & Östling, R. (2010). Political polarization and the size of government. *American Political Science Review, 104*(3), 543–65. https://doi.org/10.1017/S0003055410000262

Loidolt, S. (2018). *Phenomenology of plurality: Hannah Arendt on political intersubjectivity*. London, England: Routledge.

Low, S., Taplin, D., & Scheld, S. (2009). *Rethinking urban parks: Public space and cultural diversity*. Austin, TX: University of Texas Press.

Luckerson, V. (2015, July 9). *Here's how Facebook's news feed actually works*. TIME.com. Retrieved on januari 21, 2019 from https://time.com/collection-post/3950525/facebook-news-feed-algorithm/

Mansbach, A. (1991). Heidegger on the self, authenticity and inauthenticity. *Iyyun: The Jerusalem Philosophical Quarterly/עיון: רבעון פילוסופי*, *1*(1), 65–91.

Marres, N. (2004). Tracing the trajectories of issues, and their democratic deficits, on the web: The case of the development gateway and its doubles. *Information Technology & People, 17*(2), 124–49.

Marres, N. (2005). No issue, no public: Democratic deficits after the displacement of politics. PhD diss., University of Amsterdam.

Marres, N. (2013). Net-work is format work: Issue networks and the sites of civil society politics. In J. Dean, J. W. Anderson, & G. Lovink (Eds.), *Reformatting politics* (pp. 33–48). Oxfordshire, England: Routledge.

Martin, J. L. (2017). The human condition and the theory of action. In W. P. Baehr & P. Walsch (Eds.), *The Anthem companion to Hannah Arendt* (pp. 49–74). New York, NY: Anthem Press.

Marx, K. (1972). *The Marx-Engels reader* (Vol. 4). New York, NY: Norton.

Massey, D., & Denton, N. A. (1993). The construction of the Ghetto. Chapter 2 in *American apartheid: Segregation and the making of the underclass*. Cambridge, MA: Harvard University Press, 17–59.

McCreary, J. (2024, October 15). *Trump called Jack Smith "loony leftist"? No, image is fabricated | Fact check*. usatoday.com. https://eu.usatoday.com/story/news/factcheck/2024/10/15/trump-jack-smith-fabricated-image-fact-check/75675458007/

Mills, A. (2007). Gender and mahalle (neighborhood) space in Istanbul. *Gender, Place and Culture, 14*(3), 335–54.

Mol, A. (2021). *Eating in theory*. Durham, NC: Duke University Press.

Morozov, E. (2011). *The net delusion: How not to liberate the world*. London, England: Penguin Books.

Nichols, G. M. (2000). Primordial freedom: The authentic truth of Dasein in Heidegger's "being and time". *IWM Junior Visiting Fellows Conferences, 9*(2), 1–14.

Nietzsche, F. (1989). *On the genealogy of morals*. New York, NY: Vintage Books.

Özbank, M. (2013). *Gezi ruhu ve politik teori*. Retrieved from https://dergipark.org.tr/en/download/article-file/1731934

Özen, H. (2015). An unfinished grassroots populism: The Gezi Park protests in Turkey and their aftermath. *South European Society and Politics*, *20*(4), 533–52. https://doi.org/10.1080/13608746.2015.1099258

Özkırımlı, U. (Ed.) (2014). *The making of a protest movement in Turkey & occupy Gezi.* London, England: Palgrave Macmillan.

Papacharissi, Z. (2002). The virtual sphere: The internet as a public sphere. *New Media & Society*, *4*(1), 9–27.

Parekh, B. (1981). *Hannah Arendt and the search for a new political philosophy*. New York, NY: Springer.

Pariser, E. (2011). *The filter bubble: What the internet is hiding from you*. London, England: Penguin Books.

Passini, S. (2012). The Facebook and Twitter revolutions: Active participation in the 21st century. *Human Affairs*, *22*(3), 301–12. https://doi.org/10.2478/s13374-012-0025-0

Peeters, R. (2009a). Truth, meaning and the common world: The significance and meaning of common sense in Hannah Arendt's thought—Part one. *Ethical Perspectives*, *16*(3), 337–59.

Peeters, R. (2009b). Truth, meaning and the common world: The significance and meaning of common sense in Hannah Arendt's thought; part two. *Ethical Perspectives*, *16*(4), 411–34.

Pitkin, H. F. (1998). *The attack of the Blob: Hannah Arendt's concept of the social.* Chicago, IL: University of Chicago Press.

Raffnsøe, S. (2013). Beyond rule; trust and power as capacities. *Journal of Political Power*, *6*(2), 241–60. https://doi.org/10.1080/2158379X.2013.809216

Ramberg, B., & Gjesdal, K. (2005). Hermeneutics. Retrieved from https://stanford.library.sydney.edu.au/archives/spr2010/entries/hermeneutics/

Rao, M. B., Jongerden, J., Lemmens, P., & Ruivenkamp, G. (2015). Technological mediation and power: Postphenomenology, critical theory, and autonomist Marxism. *Philosophy & Technology*, *28*(3), 449–74.

Roodt, V. (2008). Nietzsche and/or Arendt?. In H. W. Siemens & V. Roodt (Eds.), *Nietzsche, power and politics: Rethinking Nietzsche's legacy for political thought* (pp. 411–29). New York, NY: Walter de Gruyter.

Rosenberger, R. (2017). *Callous objects: Designs against the homeless*. Minneapolis, MN: University of Minnesota Press.

Schaap, A. (2010). The politics of need. In A. Schaap, D. Celermajer, & V. Karalis (Eds.), *Power, judgment and political evil: In conversation with Hannah Arendt* (pp.157–71). Farnham: Ashgate.

Seloni, L., & Sarfati, Y. (2017). Linguistic landscape of Gezi Park protests in Turkey. *Journal of Language and Politics*, *16*(6), 782–808. https://doi.org/10.1075/jlp.15037.sel

Sunstein, C. (2007). *Republic.com 2.0*. Princeton, NJ: Princeton University Press.

Taminiaux, J. (1987). Poiesis and praxis in fundamental ontology. *Research in Phenomenology*, *17*(1), 137–69. https://www.jstor.org/stable/24654857

Taminiaux, J. (1997). *The Thracian maid and the professional thinker: Arendt and Heidegger*. Albany, NY: State University of New York Press.

Tijmes, P. (1992). The Archimedean point and eccentricity: Hannah Arendt's philosophy of science and technology. *Inquiry 35*(3), 389–406.

Till, K. E. (2005). *The new Berlin: Memory, politics, place*. Minneapolis, MN: University of Minnesota Press.

Topolski, A. (2015). *Arendt, Levinas and a politics of relationality*. Lanham, MD: Rowman & Littlefield.

Traynor, I., & Letsch, C. (2017, December 1). *Turkey divided more than ever by Erdoğan's Gezi Park crackdown*. The Guardian. Retrieved on januari 21, 2022 from https://www.theguardian.com/world/2013/jun/20/turkey-divided-erdogan-protests-crackdown

Tucker, J. (2014). *SMaPP data report: A breakout role for twitter? The role of social media in the Turkish protests*. New York University, 1–7. https://csmapnyu.org/assets/publications/2013_02_Turkish_Protests.pdf

Tufekci, Z. (2013). Pepper spray and penguins: Analysis of Turkey's social media-fueled Gezi protests. Retrieved from https://clalliance.org/blog/pepper-spray-and-penguins-analysis-of-turkey-s-social-media-fueled-gezi-protests/

Tufekci, Z. (2015, May 11). *Facebook said its algorithms do help form echo chambers. And the tech press missed it*. NOEMA. Retrieved on januari 21, 2018 from https://www.noemamag.com/facebook-said-its-algorithms-do-help-form-echo-chambers-and-the-tech-press-missed-it/

Tufekci, Z. (2017). *Twitter and tear gas: The power and fragility of networked protest*. New Haven, CT: Yale University Press.

Üngör, B. Z. (1968). Women in the Middle East and North Africa and universal suffrage. *The Annals of the American Academy of Political and Social Science*, *375*(1), 72–81.

Üşenmez, Ö., & Duman, L. (2015). Identity problems in Turkey: Alevis and AKP. *Alternatif Politika*, *7*(3), 620–44.

Veak, T. (2000). Whose technology? Whose modernity? Questioning Feenberg's questioning technology. *Science, Technology, & Human Values*, *25*(2), 226–37. https://doi.org/10.1177/016224390002500204

Verbeek, P.-P. (2005). *What things do: Philosophical reflections on technology, agency, and design*. University Park, PA: Penn State Press.

Verbeek, P.-P. (2006). Acting artifacts. In *User behavior and technology development*. New York, NY: Springer

Verbeek, P.-P. (2008a). Cyborg intentionality: Rethinking the phenomenology of human–technology relations. *Phenomenology and the Cognitive Sciences*, *7*(3), 387–95.

Verbeek, P.-P. (2008b). Obstetric ultrasound and the technological mediation of morality: A postphenomenological analysis. *Human Studies*, *31*(1), 11–26. https://doi.org/10.1007/s10746-007-9079-0

Verbeek, P.-P. (2011). *Moralizing technology: Understanding and designing the morality of things*. London, England: The University of Chicago Press.

Verbeek, P.-P. (2013). Resistance is futile: Toward a non-modern democratization of technology. *Techne: Research in Philosophy and Technology*, *17*(1), 72–92. https://doi.org/10.5840/techne20131715

Verbeek, P.-P. (2020). Politicizing postphenomenology. In G. Miller & A. Shew (Eds.), *Reimagining philosophy and technology, reinventing Ihde* (pp. 141–55). Cham, Switzerland: Springer.

Villa, D. (1995). *Arendt and Heidegger: The fate of the political*. Princeton, NJ: Princeton University Press.

Vollrath, E., & Fantel, H. (1977). Hannah Arendt and the method of political thinking. *Social Research*, *44*(1), 160–82.

Warren, J. (2004). Bios theoretikos. *The Classical Review*, *54*(2), 425–7.

Weidinger, L., Mellor, J., Rauh, M., Griffin, C., Uesato, J., Huang, P.-S., Cheng, M., Glaese, M., Balle, B., Kasirzadeh, A., Kenton, Z., Brown, S., Hawkins, W., Stepleton, T., Biles, C., Birhane, A., Haas, J., Rimell, L., Hendricks, L. A., & Gabriel, I. (2021). *Ethical and social risks of harm from language models*. arXiv. https://doi.org/10.48550/arXiv.2112.04359

Wendling, M. (2016, December 5). *The saga of "Pizzagate": The fake story that shows how conspiracy theories spread*. BBC News. https://www.bbc.com/news/blogs-trending-38156985

Wendling, M. (2024, November 3). *Voter fraud claims flood social media before US election*. BBC Breaking News, World News, US News, Sports, Business, Innovation, Climate, Culture, Travel, Video & Audio. https://www.bbc.com/news/articles/czj7eex29r3o

West, D. M. (2024, November 7). *How disinformation defined the 2024 election narrative*. Brookings. https://www.brookings.edu/articles/how-disinformation-defined-the-2024-election-narrative/

Whitehead, C., & Bozoğlu, G. (2016). Protest, bodies, and the grounds of memory: Taksim Square as "heritage site" and the 2013 Gezi protests. *Heritage & Society*, *9*(2), 111–36.

Winner, L. (1980). Do artifacts have politics? *Daedalus*, *109*(1), 121–36. https://www.jstor.org/stable/20024652

Wittingslow, R. (2020). Visions of political form: Kantian free play and urban space. Retrieved from https://contempaesthetics.org/2020/07/16/visions-of-political-form-kantian-free-play-and-urban-space/

Woolgar, S., & Cooper, G. (1999). Do artefacts have ambivalence: Moses' bridges, winner's bridges and other urban legends in S&TS. *Social Studies of Science*, *29*(3), 433–49.

Yagi, T. B. (2009). Beyond subjectivity: Kierkegaard's self and Heidegger's Dasein. *Perspectives: International Postgraduate Journal of Philosophy*, *2*(1), 60–7.

Yesil Gazete. (2013, June 7). *Gezi parkinda ne nerede* [Illustration]. Retrieved on October 5, 2016 from https://yesilgazete.org/gezi-parkinda-ne-nerede/.

Yildirim, B., & Erdem, A. (2007). Taksim Meydanının İnşası. *Tasarım+ Kuram Dergisi*, *11*(19), 95–106.

Young-Bruehl, E. (2004). *Hannah Arendt: For love of the world*. London, England: Yale University Press.

Zahavi, D. (2001). Beyond empathy: Phenomenological approaches to intersubjectivity. *Journal of Consciousness Studies*, *8*(5–7), 151–67.

Index

Page numbers in *italics* refer to figures and page numbers followed by "n" refer to end notes.

actants 8–9, 11
action 34, 83, 92, 165. *See also praxis* (action)
  characteristics 53 n.6
  definition of 32
  mediation of action 122–3
  natality, association with 165–6
  as second birth 42–3, 165, 205 n.4
  as self-disclosure 44–7
  and speech 48, 85, 101, 173–5
  spontaneity of 35, 53 n.6, 156
  as world disclosure 47–9
Akpınar, I. 133–4
alterity relation 118–19
alternative stability of technology 188–91, 195
amor mundi (love of the world) 47, 101, 156, 165
Angst (anxiety) 160, 162, 205 n.4
animal laborans 17, 24, 30
  and *homo faber* 88–91
  victory of 93–4
  worldlessness of 94–100
Aquinas, T. 63
archein (beginning) 25–7, 52 n.4
Arda, B. 182–3
Arendt, Hannah 12–14
  common sense, notion of (*see* common sense)
  conditionality, concept of 28
  Dasein as authentic political subject 162–6
  freedom, notion of 81
  The Human Condition 15, 24, 75, 82–3, 87, 92
  human distinctiveness, view on 35
  human existence, plural worldly aspect of 61
  human existence, view on 28, 59, 61, 75
  humans and things, relationship between 75
  intersubjectivity, understanding of 12
  Lectures on Kant's Political Philosophy 66
  life, view on 88
  The Life of the Mind 62
  modern technology, criticism of 86–8, 99–100
  needs, conception of 96–7
  The Origins of Totalitarianism 65
  political and social, distinction between 97–8
  political theory 12–14, 17, 23, 55, 82
  politics as *praxis* 17, 23–4
  power, view on 13
  praxis and *poiesis,* interpretation of 24–7
  On Revolution 105
  vita activa, concept of (*see vita activa* (politically active life))
  "we" for 170
  "what" and "who" of someone 155–7
  world alienation, concept of 92–4
  world-earth distinction 88–91
artificial intelligence (AI) technologies 207, 212
artworks 56, 100, 105–8, 211
autonomy 1, 6, 35
Aydin, C. 172

background relation 118–19
Balaban, O. 31
Baykan, A. 132–3
becoming, logic of 155, 180
Befindlichkeit 48, 53 n.10
being-in-the-world 59–61, 66, 157, 160, 164, 167–8
being-with-others 59–61, 164, 168
Besorgen 160
bios (human life) 52 n.3

bios politikos (life of action) 25, 32–3, 52 n.3
bios theoretikos (life of contemplation) 24–5, 52 n.2
Borren, M. 57–9, 61, 67, 72
  on common sense 63, 73–4
  condition, view of 110 n.3
  humans and world, relationship between 73
  humans as interpretative beings 58
  lived experience 57–8
  spectator, notion of 104
  subject–object dualism 110 n.1
Bozoğlu, G. 135–6

Canovan, M. 36, 76, 87–8
capitalism, origins of 95–6
care
  and authentic existence 159
  description of 159–60
  modes of 160
  of needs 81–2, 94, 96, 176–7
  for the world 47–50, 78, 94–5, 147, 156
Caro, R. A. 2
common sense 100
  based judgment 70, 104, 108
  hermeneutic circle 72–4
  preliminary and implicit understandings 73
  as *sensus communis* 66–71
  as sixth sense 61–6, 151 n.3
  sixth sense and *sensus communis,* relationship between 71–2
  technological mediation, Gezi Park 147–50
  as world building and world disclosing 18, 55, 74, 109
conditionality 28, 110 n.3
conditions 110 n.3. *See also* human condition
  human activities 29 (*see also* action; labor; work)
  natural 29
Cooper, G. 3
critical constructivism, theory of 4–5
cultural artifacts 115, 136–7
  artworks 105–8
  stories 103–5
cultural memory 114, 131, 149, 193, 209

Dahl, R. A. 6
Dasein 53 n.5, 59. *See also* being-in-the-world
  authentic/inauthentic existence 157–62
  as authentic political subject 162–6
  death, orientedness toward 159–60, 165, 205 n.4
  disclosedness 60, 159, 161, 163, 168
  moment of vision 161
  self-actualization of 163
Demirhan, K. 144
democratic interventions in technology 4–5
Dingpolitik 8–11
Disch, L. J. 104, 106–7
dominant stability 187–8, 190–1
Dossa, S. 27, 42

Eichmann, A. 171
Eken, B. 179–80
eleutheria 38
embodiment relation 118
equality 36–7, 79, 134, 147, 195
Erdoğan, T. 138–40, 144, 181–3, 198
existentialia 29–30, 52 n.5
existential/praxis-oriented postphenomenology 122–4
  composite actor 122–3
  scripts 122

Facebook, technopolitical mediation on 201–3
Fantel, H. 43–4
Feenberg, A. 2–6, 8
filter bubble 198–200
freedom 38–43
  definition of 163
  formulation of 163
  and natality 41–3
  necessity *vs.* inner freedom 38–40
  political 41, 52
  as worldly phenomenon 164
Freedom House 143–4, 152 n.20
Fürsorge 160

Gambetti, Z. 201
Gertz, N. 197, 199–200
Gezi Park, Istanbul

as ground of memory 135–6
location of park 132–3
and multistability 185–6
polis-making of 175–84
political value 130–1
Prost's master plan 133–4, 152 n.8
protests, case study on 15–16, 114
becoming, logic of 180
Gezi Commune 176–84
heterogeneous public in 130–1, 140, 150, 179, 198
Islamization, AKP government 137–8, 152 n.13
participants/protesters 140–1, 198, 201
police violence 140–2, 147, 152 n.16
social media during 130, 142–7
Taksim Pedestrianization Project 138–9
relational agency 136
and Taksim Square 132–3, 135, 191
technological mediation of common sense 147–50
technological mediation of intersubjectivity 190–2
Gjesdal, K. 73
Gohler, G. 6–7, 13, 49, 51
Gül, M. 133, 135, 138
Gurcan, E. 139

Harris, K. 1
Hatuka, T. 132–3
Haugaard, M. 6
Heidegger, M. 29–30, 48, 52 n.5, 53 n.10, 115
authentic and inauthentic existence 160–2, 164, 168
being-in-the-world, human existence as 59–61, 66, 157, 160, 164, 167–8
care, description of 159–60
Dasein, according to (*see Dasein*)
freedom, definition of 163
technology, view of 99
hermeneutic lemniscate, scheme of 18–19, 114
components 125–6, *126*
e-reader example 126–7
Gezi Park protests and 148–9
praxis and *poiesis,* interplay between 129
hermeneutic/perception-oriented postphenomenology
forms of engagement with technologies 118–19
microperception and macroperception 117–18, 151 n.3
multistability 119–22
hermeneutic phenomenology of political common sense (*see* common sense)
Dasein, Heidegger's view of 59–60
hermeneutic circle 72–4
lived experience, phenomenological analysis of 57
political phenomena 57
something-as-something, experience of 58
subject-object dualism 58, 110 n.1
world, concept of 59
hermeneutic relations 118–19
homelessness, problem of 190
homo faber 17, 24, 76–7, 85–6
attributes of 89–90
creation, types of 77
cultural products of 18, 56, 100–8, 113, 128
definition 93
durable character of things 32, 77, 90
public spaces as end products of 79
as spectator 104, 211
utility of 92–3
victory of 92
and work, activity of 89–91, 103
zoon politikon and 82
Honig, B. 44–5, 47, 155
human condition
of natality and freedom 41–3
of plurality 29, 32, 34
of worldliness 29, 31–2, 55, 74, 76, 89 (*see also* work)
human distinctness 35–7, 170–2
human existence. *See also Dasein*
Arendt's view on 28, 59, 61, 75
authentic/inauthentic modes of 157–62
plural worldly aspect of 61
human plurality 35–7, 46, 57, 172–3

human–technology relations 3–5, 118–19, 122–3
human togetherness 169, 171, 175

I-can *vs.* I-will 38–40, 165
idle talk 158, 164
Ihde, D. 115. *See also* postphenomenology of technology
  humans and technologies, relations between 118–20
  human–technology–world relations 120–2
  human–world interactions 117, 120
  material hermeneutics 124–5
  microperception/macroperception scheme 117–18, 151 n.3
  Postphenomenology 116
  technology, concept of 151 n.5, 186
İnsel, A. 180–1
intersubjectivity
  authentic and inauthentic forms of 169–70, 174 (*see also* web of relationships)
  online 212
  and political subjectivity 155–7, 167–9
  technological mediation of 190–2
  "whoness" of someone and 156–7
Inwood, M. 52 n.5, 159, 161
Islamization 137–8, 152 n.13

Jaspers, K. 56, 98–9, 115
Joerges, B. 3
judgment 66
  aesthetic 67
  common sense based 70, 104, 108
  political 67–8
  of taste 68–70
  validity of 67, 70–1

Kant 69
  Critique of Judgment 66–7, 70
Kaplan, D. M. 14–15
Karakayalí, S. 178
Kaya, A. 138
Konda Araştırma ve Danışmanlık, research company in Turkey 140–1
Kudina, O. 18–19, 113–14, 124–8, 148, 154, 185, 193–4, 204
  hermeneutic lemniscate (*see* hermeneutic lemniscate, scheme of)

labor, activity of 30, 82, 86, 88–9, 91, 94–5. *See also animal laborans*
Latour, B. 2, 8, 11
  Dingpolitik, concept of 8–10
  matter of fact and matter of concern, distinction 10–11
  nonhuman entities, importance of 9, 10
  politics of technology 11
  script, concept of 122
le bon sens (good sense) 64, 66
Lindqvist, E. 139
lived experience 57–8, 62, 72–3
Loidolt, S. 28–30, 53 n.10, 173–4
  on intersubjectivity 168–70, 174
  "who," explanation by 167–8
  on work 75
Low, S. 178

Mansbach, A. 158, 161
Marres, N. 8, 10
Marx, K., alienation theory 110 n.5
mass society 18, 56, 84, 87, 94–6, 110 n.7
material artifacts 84–5, 115, 128, 137, 151 n.5
material hermeneutics 115, 124–5
material spaces 19, 82–3, 154–5, 185, 197, 203–4, 211
material things 76, 84, 122
Merleau-Ponty, M. 62
#MeToo movement 212
microperception/macroperception scheme 117–18, 151 n.3
Minitel case 4–5
Miteinandersein (being-with-others) 61
Mol, A. 16, 130
Moses, Robert 2, 208
Moses' bridges 2–3, 6, 208–10
multistability 119–22, 154
  artifacts 187
  Gezi Park and 185–6

natality 29, 165, 205 n.4
action association with 165–6
and freedom 41–3
Nichols, G. M. 162–3
Nietzsche, F. 44–5, 172–3
Noguchi, I. 187

operation of reflection 69–70
Ostling, R. 139
other, forms of 60
Ozbank, M., *Gezi Ruhu ve Politik Teori* 146
Ozkırımlı, U. 144

Papacharissi, Z. 146
Parekh, B. 29
Pariser, E. 198–9
Peeters, R. 61–2, 64–8, 70–1
Peker, E. 139
Pitkin, H. F. 95, 110 n.6
Pizzagate conspiracy theory 211–12
plurality 35–8, 43, 51, 166
equality and distinctness 35–7
human plurality 35–6
plurality of subjectivities 167
and reality 37–8
poiesis (fabrication) 17–18, 23–4, 52, 129
in Aristotle's terminology 31–2
end products of 100, 113, 136, 150, 211
and *praxis*, intertwined relationship 55–6, 74, 109, 210–11
techne 23, 27, 31–2, 33
polis (political space) 23, 81
Gezi Park, transformation to 175–84
as model for political space 83
Plato's transformation of 24–7
politically active life. *See vita activa*
political subjectivity and intersubjectivity 155–7
authentic/inauthentic modes of human existence 157–62
Dasein as authentic political subject 162–6, 203
interconnectedness of 167–9
politics
definition of 208
Dingpolitik, notion of 8–11
features of 34–51
action 43–9
freedom 38–43
plurality 35–8
power 49–51
"in order to" relations 27, 86
as issue formation 8
performing arts and 80
as *poiesis* 17–18, 23 (*see also poiesis* (fabrication))
as *praxis* 17, 23–4 (*see also praxis* (action))
public spaces as grounds for 79
of technology 2–3, 13, 100, 207
truth and 65–6
postphenomenology of technology 18
approaches of 114–15, 117
existential/praxis-oriented 122–4
hermeneutic/perception-oriented 117–22
material hermeneutics 115
poverty 96–7
power
conceptualization of 6–7
definition 13, 49
humans and technologies, relationship between 3–5
interaction, dimension of 8, 12
of judgment 67
power over dimension 6–8, 11–12, 49, 207–8, 210
power to dimension 6–7, 49, 207
realization of 173
relationality of 7–8
of social media 198
strength, force, and authority 50
technology and 3–5
prattein (achieving) 25–7, 52 n.4
praxis (action) 17–18, 24, 26, 129
Arendtian definition of 43
Aristotle, view on 33
dependence on *poiesis* 106–7
as disclosive activity 44–7
Greek meaning 33
human action as 205 n.2
phronesis 33–4
Plato, view on 25
and *poiesis*, intertwined relationship 55–6, 74, 109, 210–11

postphenomenology, *praxis*-oriented 122–4
subjugation to *poiesis* 25–7, 86
as temporal activity 81, 102
Prost, H. 133–4, 152 n.8
public space. *See also* Gezi Park, Istanbul
definition of 185
public benches of 187–8
restrictionary modifications 189–90

Ramberg, B. 73
realms of existence, world 78
private 79, 81–2
public 78–81, 101
realness, feeling of 62–4
relations, humans and technologies 118–19
resoluteness 161, 163
revolt of dignity 180–1. *See also* Gezi Park, protests
Roodt, V. 47
Rosenberger, R. 187–90, 194

Sarfati, Y. 182
Schaap, A. 97–8
scripts 122
segregation 209–10, 213 n.1
Seloni, L. 182
sensus communis 63, 66–71
social media 15
Facebook 15, 130, 142–4, 146, 192, 196–7, 199–203
filter bubble 198–200
during Gezi Park Protests 142–7
as menace 144
power of 198
technopolitical mediation on 196–203
Twitter 15, 130, 142–6, 149, 192, 197, 199–200, 211
social realm 56, 87, 95–6, 98, 164
society
laboring 93–4
mass society 18, 56, 84, 87, 94–6, 110 n.7
plurality in 35 (*see also* plurality)
polarization in 139, 198
private individuals in 96
under totalitarian rule 172

Şoher, Ş. 133–4
Sorge 160, 162–3
space of appearance 78–80, 173–4, 193
Gezi Park as 136, 141, 149, 176, 184
peculiarity of 80–1
physical dimension 83
polis as 83–4
as temporal 81
stability of technology
dominant and alternative 187–8
multistability 119–22, 154, 185–7
subject-object dualism 57–8, 110 n.1
Sunstein, C. 199

Taminiaux, J. 25, 33
technological mediation 13–16
of common sense 129, 147–50
hermeneutic lemniscate 125–9
human-technology relations 120, 123–4, *124*
of intersubjectivity 190–2
material hermeneutics 124–5
political dimension, lack of 14–15
technology
appropriation of 126, 128, 190, 193–4, 205
in Arendt's work 86–8
artifacts, political dimension of 2–3
classical philosophy of 99
definition of 3
democratic interventions in 4–5
Heidegger, view on 99
humans and technologies, relationship between 3–5, 118–19, 122–3
Ihde's concept of 151 n.5
Minitel case 4–5
political significance of 1
political theory and philosophy of technology 15–16
and politics, inseparability of 1
postphenomenology of (*see* postphenomenology of technology)
and power 3–5
stability of (*see* stability of technology)
Tijmes, P. 76, 92–3
Till, K. E. 136, 152 n.11
Topolski, A. 60, 167, 171

totalitarianism
  Arendt's view of 65, 171–2
  and bureaucracy 172–3
  Topolski's view of 171
Trump, D. 1
truth 53 n.10
  common sense and 61–2
  factual 65–6
  rational 65
Tufekci, Z. 199–200

Verbeek, P.-P. 10, 14, 113, 117–22, 151 n.4
  on classical philosophy of technology 98–9, 115
  on cultural relation with artifact 120
  on directionality of technologies 116
  on mediation of action and perception 122–3
  technological mediation theory 5, 14–16, 115, 120–1, 123–4, *124*
Villa, D. 45, 59–60, 76, 90, 160, 163
vita activa (politically active life) 25, 28–30, 52 n.3, 86, 94
  action 32–4
  hierarchy of 29–30, 91, 100
  labor 30
  modes of 17, 24
  work 31–2
vita contemplativa (life of contemplation) 25, 52 n.2
Vollrath, E. 43–4

Walz, T. 1
"we," authentic form of 169–70, 173, 179–84
Weber, M. 95
web of relationships 169–70
  authentic 173–5
  inauthentic 170–3
"what" and "who" of someone 45–6, 155–7, 167–8, 170, 174–5
Whitehead, C. 135–6
Winner, L. 2–3, 6, 8, 207–9
  politics, definition of 208
with-world *(Mitwelt)* 59, 157
Wittingslow, R. 187, 191
Woolgar, S. 3
work, activity of 31–2, 55, 74–5, 89–91. *See also poiesis* (fabrication)
  for Arendt 89–91, 99
  human-made things and 82, 86
  Loidolt on 75
world
  alienation 92–4, 98
  and earth, distinction between 88–91
  homo faber, creations of 77 (*see also homo faber*)
  as material construct 83–6
  private realm of existence 79, 81–2
  public realm of existence 78–81, 101
  realms of existence 78
worldlessness and social realm 56, 94–100, 109
worldliness 29–31
  common world and 74–77
  and social realm 94–100

Yaka, Ö. 178

Zahavi, D. 168
zoon politikon 17, 24, 32, 76–8, 86, 104
  and *homo faber* 82, 100, 129
  political realm of 82
  space of appearance 78

www.ingramcontent.com/pod-product-compliance
Ingram Content Group UK Ltd.
Pitfield, Milton Keynes, MK11 3LW, UK
UKHW020213290926
15470UKWH00001B/158